# MATHEMATICAL STATISTICS AND PROBABILITY THEORY

Volume B

Statistical Inference and Methods

# MATHEMATICAL STATISTICS AND PROBABILITY THEORY

## Volume B

## Statistical Inference and Methods

*Proceedings of the 6th Pannonian Symposium on Mathematical Statistics, Bad Tatzmannsdorf, Austria, September 14-20, 1986*

Edited by

P. BAUER
*University of Cologne, F.R.G.*

F. KONECNY
*University of Agriculture, Vienna, Austria*

and

W. WERTZ
*Technical University, Vienna, Austria*

D. REIDEL PUBLISHING COMPANY

A MEMBER OF THE KLUWER ACADEMIC PUBLISHERS GROUP

DORDRECHT / BOSTON / LANCASTER / TOKYO

**Library of Congress Cataloging in Publication Data**

CIP

Mathematical Statistics of probability theory.

Papers presented at the sixth Pannonian Symposium on Mathematical Statistics held in Bad Tatzmannsdorf, Sept. 14–20, 1986.

Includes indexes.

Contents: v. A. Theoretical aspects / edited by M. L. Puri, P. Révész, and W. Wertz — v. B. Statistical inference and methods / edited by P. Bauer, F. Konecny, and W. Wertz.

1. Mathematical statistics—Congresses. 2. Probabilities—Congresses. I. Puri, Madan Lal. II. Pannonian Symposium on Mathematical Statistics (6th: 1986: Bad Tatzmannsdorf, Austria)

QA276.A1M27 1987 519.5 87–24338

ISBN-13: 978-94-010-8259-4 e-ISBN-13: 978-94-009-3965-3

DOI: 10.1007/978-94-009-3965-3

---

Published by D. Reidel Publishing Company,
P.O. Box 17, 3300 AA Dordrecht, Holland.

Sold and distributed in the U.S.A. and Canada
by Kluwer Academic Publishers,
101 Philip Drive, Assinippi Park, Norwell, MA 02061, U.S.A.

In all other countries, sold and distributed
by Kluwer Academic Publishers Group,
P.O. Box 322, 3300 AH Dordrecht, Holland.


Softcover reprint of the hardcover 1st edition 1987

CONTENTS

## PREFACE

The past several years have seen the creation and extension of a very conclusive theory of statistics and probability. Many of the research workers who have been concerned with both probability and statistics felt the need for meetings that provide an opportunity for personal contacts among scholars whose fields of specialization cover broad spectra in both statistics and probability: to discuss major open problems and new solutions, and to provide encouragement for further research through the lectures of carefully selected scholars, moreover to introduce to younger colleagues the latest research techniques and thus to stimulate their interest in research.

To meet these goals, the series of Pannonian Symposia on Mathematical Statistics was organized, beginning in the year 1979: the first, second and fourth one in Bad Tatzmannsdorf, Burgenland, Austria, the third and fifth in Visegrád, Hungary. The Sixth Pannonian Symposium was held in Bad Tatzmannsdorf again, in the time between 14 and 20 September 1986, under the auspices of Dr.Heinz FISCHER, Federal Minister of Science and Research, Theodor KERY, President of the State Government of Burgenland, Dr.Franz SAUERZOPF, Vice-President of the State Government of Burgenland and Dr.Josef SCHMIDL, President of the Austrian Statistical Central Office. The members of the Honorary Committee were Pál ERDÖS, Władisław ORLICZ, Pál RÉVÉSZ, Leopold SCHMETTERER and István VINCZE; those of the Organizing Committee were Wilfried GROSSMANN (University of Vienna), Franz KONECNY (University of Agriculture of Vienna) and, as the chairman, Wolfgang WERTZ (Technical University of Vienna).

About 160 scholars from 17 countries participated in this conference, a particularly large number of them came from Hungary, Poland and Germany, but more distant countries were well-represented, too, such as The Netherlands, Spain and Portugalia; moreover there were several participants from the United States of America, Canada, Israel and the Republic of South Africa.

The scientific program of the Sixth Pannonian Symposium on Mathematical Statistics covered more than 100 contributions, most of them in the form of contributed lectures, a few of them in the framework of a poster session. The four specially invited plenary lectures were delivered by Luc DEVROYE (Montréal), Herbert HEYER (Tübingen), Petr MANDL (Praha) and Madan L.PURI (Bloomington). There was a rather broad range of the topics, including probability theory, theory of stochastic processes, the mathematical foundations of statistics, decision theory, statistical methods and some applications.

A selection of the contributions of the conference is published in these proceedings, consisting of two volumes. Whereas this book contains papers emphasizing the development of statistical and probabilistic methods, the other volume, with the subtitle "Theoretical Aspects", includes primarily contributions concerned with the mathematical foundations of statistics and probability theory (a list of the contents of this volume can be found on p.261). It has been the aim of the editors to publish new and significant results; the assistance of numerous referees constituted an indispensible help in approaching this objective - the editors wish to express their deep gratitude to all the referees; they are listed below. Despite of the careful redaction of the volume, the responsibility for the manuscripts remains with the authors.

Roughly speaking, the papers of this volume appertain four main topics: probability and stochastic processes, testing hypotheses, estimation and applications. The *probabilistic* articles have obvious applications in statistics or even explicitely refer to them; to the first group belong the papers by Erdös&Révész (on random walks; several interesting open problems are formulated), Glänzel (characterization theorems), Gyires (linear prediction), Móri (limit theorems for waiting times) and Stadje (stopping of processes); with the second one rank the articles by Athayde&Gomes (extreme value limit theorems applied to testing problems) and by Ignatov&Kaishev (certain distributions applied to contingency tables). Three papers deal with *testing statistical hypotheses*: Bajorski&Ledwina (rank tests), Drost (chi-square type tests) and Práškova&Ratajová (Bayesian analysis of contingency tables, in particular the construction of credible intervals). Various *estimation problems* are considered: density estimation for dependent samples by Deddens&Peligrad&Yang, estimators based on censored data by Ferenstein, Schick&Susarla, parameter estimation after certain Box-Cox-transformations by Rukhin and sequential estimation in stochastic processes by Pruscha; González Manteiga&Vilar Fernández use nonparametric criteria for estimating parameters of time series. *Methods and applications* are dealt with by Banjević&Nedeljković (technical application), Gupta&Liang (selection procedures), Krzyśko&Wachowiak (classification), Malisić (time series models) and Weron&Weron (use of stable distributions in relaxation problems). Mandl surveys certain connections between statistics and control theory.

The organization of the Sixth Pannonian Symposium on Mathematical Statistics was made possible by the valuable help of many institutions and individuals. The organizers take the opportunity to express their thanks, in particular, to the following institutions: the State Government of Burgenland (Departments of Official Statistics, of Affairs of Communes and of Tourist Trade), the Federal Ministry of Science and Research, the Austrian Statistical Society, the Creditanstalt, the Volksbank Oberwart, the Raiffeisenverband Burgenland, the Local Government Bad Tatzmannsdorf, the Kurbad Tatzmannsdorf AG and the Spa Commission of Bad Tatzmannsdorf. The interest of the Authorities in the conference has been emphasized by the attendance of numerous representatives of public life at the opening ceremony of the symposium; the President of the State Government of Burgenland honoured the congress by opening it himself.

Last not least, cordial thanks are due to the ladies who helped in the local organization and in mastering the extensive paperwork and typing.

Bad Tatzmannsdorf,
April 1987

Wolfgang Wertz

## ACKNOWLEDGEMENT

We express our deepest gratitude to the following referees, who gave us indispenable advice for the editorial process. They helped us in the selection of the papers published in the two proceedings volumes on the Sixth Pannonian Symposium on Mathematical Statistics; their constructive criticism and numerous valuable suggestions to the authors lead to a considerable improvement of several manuscripts.

The editors

## LIST OF REFEREES

James ALBERT (Bowling Green, USA, and Southampton, UK)
Jiří ANDĚL (Praha, Czechoslovakia)
Per Kragh ANDERSEN (Kobenhaven, Denmark)
Tadeusz BEDNARSKI (Wrocław, Poland)
Patrick BILLINGSLEY (Chicago, USA)
Denis BOSQ (Villeneuve d'Ascq, France)
Richard BRADLEY (Bloomington, USA)
Włodzimierz BRYC (Cincinnati, USA)
Raymond J.CARROLL (Chapel Hill, USA)
S.D.CHATTERJI (Lausanne, Switzerland)
Yuan S.CHOW (New York, USA)
Arthur COHEN (New Brunswick, USA)
Endre CSÁKI (Budapest, Hungary)
Miklós CSÖRGŐ (Ottawa, Canada)
Paul DEHEUVELS (Paris, France)
Manfred DENKER (Göttingen, FR Germany)
Luc DEVROYE (Montreal, Canada)
Paul DOUKHAN (Orsay, France)
Edward J.DUDEWICZ (Syracuse, USA)
John H.EINMAHL (Maastricht, Netherlands)
Jürgen FRANZ (Dresden, German Dem.Rep.)
János GALAMBOS (Philadelphia, USA)
Erhard GODEHARD (Düsseldorf, FR Germany)
Friedrich GÖTZE (Bielefeld, FR Germany)
Karl GRILL (Wien, Austria)
Wilfried GROSSMANN (Wien, Austria)
Shanti S.GUPTA (West Lafayette, USA)
László GYÖRFI (Budapest, Hungary)
Jürgen HAFNER (Wien, Austria)
Marc HALLIN (Bruxelles, Belgium)
D.J.HAND (London, UK)

Wilfried HAZOD (Dortmund, FR Germany)
Bernard HEINKEL (Strasbourg, France)
Herbert HEYER (Tübingen, FR Germany)
Omar HIJAB (Philadelphia, USA)
Albrecht IRLE (Kiel, FR Germany)
Jana JUREČKOVÁ (Praha, Czechoslovakia)
Michał KARONSKI (Poznań, Poland)
Gerhard KELLER (Heidelberg, FR Germany)
Jacek KORONACKI (Warszawa, Poland)
Samuel KOTZ (Columbus, USA)
Andrzej KOZEK (Wrocław, Poland)
E.KREMER (Hamburg, FR Germany)
A.J.LAWRENCE (Birmingham, UK)
Alexander LEITSCH (Delaware, USA)
Margit LÉNÁRD (Debrecen, Hungary)
Antonín LEŠANOVSKÝ (Praha, Czechoslovakia)
Dennis V.LINDLEY (Minehead, UK)
Harald LUSCHGY (Münster, FR Germany)
James LYNCH (Columbia, USA)
James B.MAC QUEEN (Los Angeles, USA)
Ryszard MAGEIRA (Wrocław, Poland)
David M.MASON (Newark, USA)
Jochen MAU (Tübingen, FR Germany)
Klaus J.MISCKE (Chicago, USA)
Itrel MONROE (Fayetteville, USA)
David S.MOORE (West Lafayette, USA)
Tamás MÓRI (Budapest, Hungary)
Ferenc MÓRICZ (Szeged, Hungary, and Syracuse, USA)
Harald NIEDERREITER (Wien, Austria)
Jacobus OOSTERHOFF (Amsterdam, Netherlands)
Magda PELIGRAD (Cincinnati, USA)
Walter PHILIPP (Urbana, USA)
Detlef PLACHKY (Münster, FR Germany)
Benedikt M.PÖTSCHER (Wien, Austria, and New Haven, USA)
Prem S.PURI (New Delhi, India)
Ronald PYKE (Seattle, USA)
Lidia REJTŐ (Budapest, Hungary)
H.RINNE (Gießen, FR Germany)
Vijay K.ROHATGI (Bowling Green, USA)
Günter ROTHE (Mannheim, FR Germany)
Andrew RUKHIN (Amherst, USA)
Wolfgang RUPPERT (Wien, Austria)
Zdzisław RYCHLIK (Lublin, Poland)
Stephen M.SAMUELS (West Lafayette, USA)
Klaus D.SCHMIDT (Mannheim, FR Germany)
Norbert SCHMITZ (Münster, FR Germany)
Claus-Peter SCHNORR (Frankfurt, FR Germany)
Pranab K.SEN (Chapel Hill, USA)
Wolfgang SENDLER (Trier, FR Germany)
Galen R.SHORACK (Seattle, USA)
Robert H.SHUMWAY (Davis, USA)

Bernard W.SILVERMAN (Bath, UK)
Richard L.SMITH (Surrey, UK)
Michael SØRENSEN (Aarhus, Denmark)
Valeri STEFANOV (Sofia, Bulgaria)
Josef STEINEBACH (Marburg, FR Germany)
Larry STOCKMEYER (San Jose, USA)
Helmut STRASSER (Bayreuth, FR Germany)
Harald STRELEC (Wien, Austria)
Louis SUCHESTON (Columbus, USA)
V.SUSARLA (Binghamton, USA)
Domokos SZÁSZ (Budapest, Hungary)
Gábor J.SZÉKELY (Budapest, Hungary)
Dominik SZYNAL (Lublin, Poland)
J.A.TAWN (Surrey, UK)
Erik TORGERSEN (Oslo, Norway)
Gábor TUSNÁDY (Budapest, Hungary)
Reinhard VIERTL (Wien, Austria)
Grace WAHBA (Madison, USA)
Ishak WEISSMAN (Davis, USA)
Hans-Joachim WERNER (Bonn, FR Germany)
Aleksander WERON (Wrocław, Poland)
Jan C.WILLEMS (Groningen, Netherlands)
Hermann WITTING (Freiburg im Breisgau, FR Germany)
Michael B.WOODROOFE (Ann Arbor, USA)
Franz ZIEGLER (Wien, Austria)

# MULTIVARIATE EXTREMAL MODELS UNDER NON-CLASSICAL SITUATIONS

Emilia Athayde and M. Ivette Gomes
Faculty of Sciences of Lisbon
Center of Statistics and Applications (I.N.I.C.)
58, Rua da Escola Politécnica
1294 Lisboa Codex - Portugal

ABSTRACT. The limiting distribution of top order statistics in a non-classical set-up, where the independence structure remains valid, is reviewed in this paper. We essentially place ourselves under Mejzler's hypothesis - independent $X_k$'s with distribution function $F_k(x)$, $k \geq 1$, satisfying the uniformity condition for the maximum. Notice that the results presented are obviously valid not only on Mejzler's $\mathbf{M}_1$ class, but also on refinements $\mathbf{M}_r$, $r>1$, of Mejzler's class and in $\mathbf{M}_\infty = \bigcap_{r \geq 1} \mathbf{M}_r$, a non-trivial extension of the class S of max-stable distributions. Generalizing the multivariate GEV model, other multivariate extremal models based on functions H(x) belonging to $\mathbf{M}_1$ (or to $\mathbf{M}_\infty$) are introduced and inference techniques are developed for a multivariate extremal Pareto model.

## 1. INTRODUCTION AND PRELIMINARIES

For sequences of independent, identically distributed (i.i.d.) random variables (r.v.'s) the non-degenerate limiting structure, whenever it exists, of the normalized top i order statistics (o.s.), i a fixed integer, is well-known and characterized by the joint probability density function (p.d.f.)

$$h(z_1,\ldots,z_i) = g(z_i) \prod_{j=1}^{i-1} \{g(z_j)/G(z_j)\}, z_1>\ldots>z_i \tag{1}$$

where $G(z) = G_\theta(z)$ is in the class S of max-stable distribution functions (d.f.'s), often called Generalized Extreme Value (GEV) d.f.'s, i.e.

$$G_\theta(z) = \begin{cases} \exp(-(1-\theta z)^{1/\theta}), \ 1-\theta z>0, \ z \in \mathbb{R} & \text{if } \theta \neq 0 \\ \exp(-\exp(-z)), \ z \in \mathbb{R} & \text{if } \theta=0 \end{cases} \tag{2}$$



*P. Bauer et al. (eds.), Mathematical Statistics and Probability Theory, Vol. B, 1–9.*

$g_\theta(z) = \partial G_\theta(z)/\partial z$.

If we drop the hypothesis of identical distribution — a more common set-up in applications — , and deal with sequences $\{Y_n\}_{n\geq 1}$ of r.v.'s whose associated sequence of partial maxima $\{M_n^{(1)} = \max_{1\leq j\leq n} Y_j\}_{n\geq 1}$, suitably normalized, converges weakly, as $n\to\infty$, to a r.v. in Mejzler's class $\mathbf{M}_1$ [Mejzler, 1956], the limiting structure of the top i o.s., detailed in section 2, is still a multivariate extremal vector with p.d.f. given by (1), but with $G\varepsilon\,\mathbf{M}_1$. Notice that this is just a corollary of the result of Weissman (1975), expressed here in a slightly different context. If we work instead with refinements $\mathbf{M}_r$, r>1, of Mejzler's class $\mathbf{M}_1$, or with $\mathbf{M}_\infty = \bigcap_{r\geq 1} \mathbf{M}_r$ [Graça Martins and Pestana, 1985], analogous results are obtained. Notice that $\mathbf{M}_\infty$ is the smallest class containing the class S of max-stable d.f.'s (2) that is closed under pointwise products and limits. This class seems thus to provide a very general framework for the study of sample maxima, justifying several models put forward by statistical users [Gomes and Pestana, 1985], the same happening to its multivariate generalizations presented here.

This limiting result is thus the probabilistic background for the introduction of a multivariate extremal $\mathbf{M}_1$ model to analyse the set of the largest observations available, in order to infer tail properties. This model is obviously more general and more realistic than the multivariate extremal GEV model, introduced first, in a slightly different context, by Pickands (1975) and worked out by several authors [Weissman, 1978; Smith, 1984]. Mainly for climatological data, where the i.d. hypothesis fails, a multivariate extremal $\mathbf{M}_1$ model, or at least a multivariate extremal $\mathbf{M}_\infty$ model, has often to be called for.

Among the members of the class $\mathbf{M}_1$ we consider, in sections 3 and 4, the Pareto d.f.'s

$$H_\theta(z) = \begin{cases} (1+\theta z)^{1/\theta}, \; 1+\theta z>0, \; z<0 & \text{if } \theta\neq 0 \\ \exp(z), \; z<0 & \text{if } \theta=0 \end{cases} \tag{3}$$

and the multivariate extremal model $(X_1,X_2,\ldots,X_{m+1})$, with p.d.f. given by (1), i replaced by m+1 and G(z) replaced by $H_\theta((x-\lambda)/\delta)$, $\lambda\in\mathbb{R}$ and $\delta\in\mathbb{R}^+$ unknown location and scale parameters respectively to be estimated from the sample. We shall call it a multivariate extremal $H_\theta$ model (notice that $H_\theta\,\varepsilon\,\mathbf{M}_\infty$ if $\theta\geq 0$). Since inference techniques in this model are more easily developed for the particular case $\theta=0$, we shall deal here with discrimination among these models, with $\theta=0$ playing a central and eminent role. With this statistical choice problem in mind, we shall deal with Gumbel statistic

$$G_{m+1} = \{X_1-X_{r_m}\}/\{X_{r_m}-X_{m+1}\}, \quad r_m=[(m+1)/2]+1 \tag{4}$$

for testing $H_o:\theta=0$, in a multivariate extremal $H_\theta$ model, versus one-sided or two-sided alternatives, [y] denoting, as usual, the greatest integer smaller than y. This statistic, although suggested to us by means of heuristic reasons turned out to be powerful for testing $H_0:\theta=0$ in a univariate GEV($\theta$) model [van Montfort and Gomes, 1985, and references therein] and in a multivariate GEV($\theta$) model [Gomes and Alpuim,1986]. In this same context, an analogue of the Locally Most Powerful (LMP) test statistic

$$\tilde{L}_{m+1} = \sum_{j=1}^{m} \{X_1-X_j\}/\{X_1-X_{m+1}\} \tag{5}$$

is also considered.

In section 3 of this paper we derive (5) and distributional properties of $G_{m+1}$ and $\tilde{L}_{m+1}$, under the validity of the null hypothesis $H_0:\theta=0$. Notice that, although the limiting result, in section 2, is obtained for i fixed, in applications, the model remains valid for reasonably large values of sample size, and asymptotic properties of statistics related to the model may thus be called for.

Finally, in section 4, we compare the power functions of the tests based on the statistics (4) and (5).

## 2. LIMITING STRUCTURE OF TOP ORDER STATISTICS IN A NON-CLASSICAL SET-UP

Let $\{Y_n\}_{n\geq 1}$ denote a sequence of independent r.v.'s and let $F_j(.)$ be the d.f. of $Y_j$, $j\geq 1$. Let us denote by $M_n^{(k)}$ the k-th maximum value of $\{Y_1,\ldots,Y_n\}$, $1\leq k\leq n$, and assume further the validity of the uniformity condition for the maximum [Mejzler, 1956], and for a suitable sequence $\{u_n\}_{n\geq 1}$ of real numbers, i.e.

$$\lim_{n\to\infty} \min_{1\leq j\leq n} F_j(u_n) = 1 \tag{6}$$

The limiting structure of $(M_n^{(1)},\ldots,M_n^{(i)})$, i a fixed integer, may thus be derived, analogously to what has been done in an i.i.d. set-up [Leadbetter, Lindgrèn and Rootzèn, 1983], from the limiting structure of $M_n^{(1)}$, obtained by Mejzler.

As a direct corollary of the results on convergence of Bernoulli point processes [Serfozo, 1986] to Poisson point processes, as the probabilities of success are small, we obtain

Theorem 1. Under the conditions stated before, assume that $u_n^{(j)}=u_n^{(j)}(\tau_j)$, are such that

$$1-\frac{1}{n}\sum_{k=1}^{n} F_k(u_n^{(j)}(\tau_j)) = \tau_j/n + o(\tfrac{1}{n}), \text{as } n\to\infty, 1\leq j\leq i, \tag{7}$$

that condition (6) for $u_n^{(j)}(\tau_j)$, $1\leq j\leq i$, $\tau_1<\tau_2<\ldots<\tau_i$ is valid, and let $S_n^{(j)}$ be the number of exceedances of $u_n^{(j)}$ by $\{Y_1,\ldots,Y_n\}$, $1\leq j\leq i$, $n\geq 1$.

Then

$$\lim_{n\to\infty} P[\bigcap_{j=1}^{i}\{S_n^{(j)}=k_j\}] = \begin{cases} e^{-\tau_i}\dfrac{\tau_1^{k_1}}{k_1!}\displaystyle\prod_{j=1}^{i-1}\dfrac{(\tau_{j+1}-\tau_j)^{k_{j+1}-k_j}}{(k_{j+1}-k_j)!} & \text{if } k_1\le\ldots\le k_i \\ 0 & \text{otherwise} \end{cases} \tag{8}$$

i.e., the r.v.'s $S_n^{(j-1,j)} = \#\{k: 1\le k\le n,\ u_n^{(j)}<Y_k\le u_n^{(j-1)}\}$, $1\le j\le i$, $u_n^{(0)}=+\infty$, are asymptotically independent Poisson r.v.'s with mean value $\tau_j-\tau_{j-1}$, $1\le j\le i$, $\tau_0=0$.

Notice that different versions of this result are used by Weissman (1975) and Hall (1978). Indeed our theorem 1 corresponds to a re-statement of theorem 1 of Weissman (1975), with t=1, assuming additionally the uniformity condition (6), which is there derived from a broader hypothesis on extremal processes.

Since

$$P[\bigcap_{j=1}^{i}\{M_n^{(j)}\le u_n^{(j)}\}] = P[\bigcap_{j=1}^{i}\{S_n^{(j)}<j\}] \tag{9}$$

we have

Theorem 2. Under the conditions stated before, assume that there exists sequences of real constants $\{a_n\}_{n\ge1}$ $(a_n>0)$, $\{b_n\}_{n\ge1}$ and a non-degenerate continuous d.f. H(z) such that

$$\lim_{n\to\infty} P[M_n^{(1)}\le a_n z+b_n] = H(z),\quad z\in\mathbb{R} \tag{10}$$

being additionally valid the uniformity condition for the maximum (6), and for sequences $u_n=a_n z+b_n$, $z\in\mathbb{R}$, $n\ge1$. Then, for any fixed integer i there exists a non-degenerate i-variate d.f. $H(z_1,\ldots,z_i)$ such that

$$\lim_{n\to\infty} P[\bigcap_{j=1}^{i}\{M_n^{(j)}\le a_n z_j+b_n\}] = H(z_1,\ldots,z_i) \tag{11}$$

to which corresponds, whenever h(z)=H'(z) exists, a p.d.f. given by (1), g(.) and G(.) replaced by h(.) and H(.) respectively.

Notice that under the context stated before, also this theorem may be derived from theorem 3 of Weissman (1975) (taking there q=1, $t_q=1$).

In the set-up considered, with no further restrictions, H(z) is in Mejzler's class $\mathbf{M}_1$ of d.f.'s such that (cf. Galambos (1978), p. 181) either

(i) $-\log\{H(x)\}$ convex

or

(ii) $R=\sup\{x: H(x)<1\}$ finite and $-\log\{H(R-\exp(-x))\}$ convex.

Notice however that the results in theorem 2 remain valid if we work with more restrictive sequences of r.v.'s satisfying the uniformity condition for the maximum, like set-ups that lead us to the refinements $\mathbf{M}_r$, r>1, of Mejzler's class $\mathbf{M}_1 \supset \mathbf{M}_2 \supset \dots$ introduced by Graça Martins and Pestana (1985), or to $\mathbf{M}_\infty = \bigcap_{r\geq 1} \mathbf{M}_r$. Classes $\mathbf{M}_r$, r>1, and $\mathbf{M}_\infty$ are characterized like $\mathbf{M}_1$ by (i) and (ii), with convexity replaced by monotonicity of order r, r>1, and complete monotonicity respectively.

Notice also that $H_\theta$ defined in (3) is a member of $\mathbf{M}_1$ for every $\theta \in \mathbb{R}$. More than that: $H_\theta$ belongs to $\mathbf{M}_\infty$ if and only if $\theta \geq 0$.

## 3. DISTRIBUTIONAL BEHAVIOUR OF TEST STATISTICS, UNDER A MULTIVARIATE EXTREMAL $H_0$ MODEL

We shall consider here the multivariate extremal $H_\theta$ model $\underset{\sim}{X}=(X_1,\dots,X_{m+1})$ where $\underset{\sim}{Z} = (Z_j=(X_j-\lambda)/\delta,\ 1\leq j\leq m+1)$, $\lambda \in \mathbb{R}$, $\delta \in \mathbb{R}^+$, has a p.d.f. $h_\theta(z_1,\dots,z_{m+1})$ given by (1), i=m+1, and G(.) replaced by the Pareto d.f. $H_\theta(.)$ in (3). As mentioned before, our interest lies in testing $H_o$: $\theta=0$ versus suitable one-sided or two-sided alternatives, and we first use Gumbel statistic $G_{m+1}$ defined by (4).

Notice that $G_{m+1}$ is invariant under location and scale transformations, i. e., $G_{m+1} = G_{m+1}(\underset{\sim}{X}) = G_{m+1}(\underset{\sim}{Z})$. Under $H_o$:$\theta=0$, $Z_j-Z_{j+1}$, $1\leq j\leq m$, are independent exponential r.v.'s, and consequently $G_{m+1}$ is, for m even, the quotient of two Gamma(m/2) independent r.v.'s, and for m odd, the quotient of independent Gamma((m+1)/2) and Gamma((m-1)/2) r.v.'s. Consequently $G_{m+1}$ is distributed as F(m,m), when m is even, and as (m+1)F(m+1,m-1)/(m-1) when m is odd. $F(\nu_1,\nu_2)$ denotes, as usual, the F-distribution with parameters $(\nu_1,\nu_2)$. We consequently consider the test statistic

$$G^*_{m+1} = \sqrt{m}\,\{G_{m+1}-1\}/2 \tag{12}$$

which is asymptotically, as $m\to\infty$, a standard normal r.v..

For small m, tables of the F-distribution may thus be used to obtain percentage points of $G^*_{m+1} \mid H_o$:$\theta=0$ in the multivariate extremal $H_0$ model, both for one-sided and two-sided alternatives.

In the same context of statistical choice in a multivariate extremal $H_\theta$ model, an analogue of the LMP test statistic is considered. Indeed, in the standard model $\underset{\sim}{Z}=(Z_1,\dots,Z_{m+1})$, the LMP test statistic for $H_o$:$\theta=0$ is, asymptotically,

$$L_{m+1}(\underset{\sim}{Z}) = \left.\frac{\partial \log h_\theta(Z_1,\dots,Z_{m+1})}{\partial \theta}\right|_{\theta=0} = -\sum_{j=1}^{m+1} Z_j - Z^2_{m+1}/2 \tag{13}$$

for both one-sided or two-sided alternatives.

When working with the general model $\underset{\sim}{X}=(X_1>X_2>\ldots>X_{m+1})$, we consider, as usual, the test statistic

$$\hat{L}_{m+1}(\underset{\sim}{X}) = L_{m+1}((\underset{\sim}{X}-\hat{\lambda}_0\ \underset{\sim}{1})/\hat{\delta}_0) \tag{14}$$

where $(\hat{\lambda}_0,\hat{\delta}_0)$ are the maximum likelihood estimators of the unknown parameters $(\lambda,\delta)$ under $H_0:\theta=0$, $\underset{\sim}{1}$ being a column vector with all its components equal to one. Since we have

$$\hat{\lambda}_0 = X_1 \ ; \ \hat{\delta}_0 = (X_1-X_{m+1})/(m+1) \tag{15}$$

we finally obtain, after a few manipulations, $\tilde{L}_{m+1}$, given by (5), an equivalent analogue of the LMP test statistic.

Since, under $H_0:\theta=0$, $\{Z_j-Z_{m+1}\}/\{Z_1-Z_{m+1}\}=\{\sum_{k=j}^{m} V_k\}/\{\sum_{k=1}^{m} V_k\}$, $2\leq j\leq m$, $\{V_k\}_{1\leq k\leq m}$ i.i.d. exponential r.v.'s, are the descending order statistics associated to a sample $U_k$, $1\leq k\leq m-1$, of i.i.d. Uniform(0,1) r.v.'s, we have the distributional identity

$$\tilde{L}_{m+1} = (m-1) - \sum_{k=1}^{m-1} U_k \tag{16}$$

The test statistic considered here is thus

$$L^*_{m+1} = \sqrt{12(m-1)}\ \{ \frac{1}{m-1}\ \tilde{L}_{m+1} - 1/2 \} \tag{17}$$

which is asymptotically, as $m\to\infty$, and under $H_0:\theta=0$, a standard normal r.v..

Since the sum of uniform r.v.'s in (16) converges fast to the normal distribution, the standard normal percentiles may be used to a very good accuracy, when dealing with the test statistic (17), even for quite small m.

## 4. COMPARISON OF TEST STATISTICS

Simulation of the multivariate extremal $H_\theta$ model, $H_\theta$ given by (3), is straightforward: from a set $\{R_i\}_{i\geq 1}$ of pseudo random numbers in (0,1), we compute, for $1\leq j\leq m+1$

$$Z_j = H_\theta^{-1}(\prod_{k=1}^{j} R_k) = \begin{cases} \{(\prod_{k=1}^{j} R_k)^\theta - 1\}/\theta & \text{if } \theta\neq 0 \\ \sum_{k=1}^{j} \log(R_k) & \text{if } \theta=0 \end{cases}$$

Comparison of test statistics is thus made by simulation. In table I we present results regarding the power functions of the statistical choice tests based on $G_{m+1}$ and $\tilde{L}_{m+1}$ for m=20 and m=60 and for testing $H_o: \theta=0$ versus $H_1: \theta\neq 0$ in the multivariate extremal $H_\theta$ model. For each value of $\theta$ and for each test statistic we give the simulated power of that test statistic. The number of runs in each simulation was chosen such that the standard errors associated to powers are smaller than .005. Blank entries correspond to simulated powers higher than .995.

Figure 1 is a visual representation of table I, m=20.

Notice that, contrary to what happened in a multivariate GEV($\theta$) model, the power function of the LMP test statistic turns out to be, uniformly over $\theta \in \mathbb{R}$, higher than the power function of the Gumbel statistic, for testing $H_o:\theta=0$ in a multivariate extremal $H_\theta$ model. It is natural that, asymptotically, the same happens, since $\tilde{L}_{m+1}$ was built according to an 'optimal' asymptotic criterion whereas $G_{m+1}$ was merely based on heuristic reasons. Asymptotic power of these statistics, for testing $H_o:\theta=0$ in a multivariate extremal $H_\theta$ model, is under investigation. It is however worth mentioning that the naïve and simple statistic $G_{m+1}$ is practically (almost) as good as the LMP test statistic.

TABLE I. Comparative power functions of $G_{m+1}$ and $\tilde{L}_{m+1}$ for m=20,60 and at a significance level $\alpha$=.05

| | m = 20 | | m = 60 | |
|---|---|---|---|---|
| $\theta$ | $G_{m+1}$ | $\tilde{L}_{m+1}$ | $G_{m+1}$ | $\tilde{L}_{m+1}$ |
| -.30 | .99 | | | |
| -.25 | .98 | .99 | | |
| -.20 | .94 | .97 | | |
| -.15 | .84 | .89 | | |
| -.125 | .73 | .80 | | |
| -.1 | .57 | .63 | | |
| -.075 | .38 | .44 | | |
| -.05 | .21 | .24 | | |
| -.025 | .09 | .10 | .80 | .88 |
| .025 | .09 | .10 | .81 | .89 |
| .05 | .20 | .24 | | |
| .075 | .38 | .45 | | |
| .1 | .56 | .63 | | |
| .125 | .72 | .79 | | |
| .15 | .83 | .88 | | |
| .20 | .93 | .96 | | |
| .25 | .98 | .99 | | |
| .30 | .99 | | | |

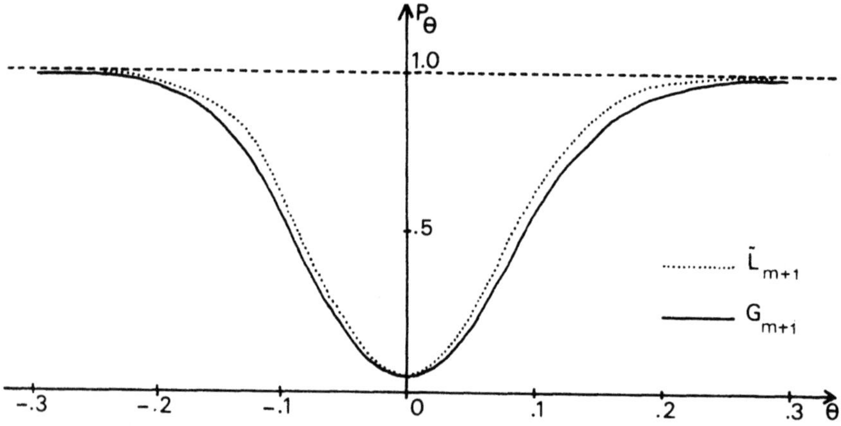


Figure 1. Power functions of $G_{m+1}$ and $\tilde{L}_{m+1}$, m=20, α=.05

ACKNOWLEDGEMENTS

The authors are grateful to the anonymous referees for helpful comments.

REFERENCES

Galambos, J. (1978). The Asymptotic Theory of Extreme Order Statistics. Wiley, New York.

Gomes, M.I. and Alpuim, M.T. (1986). 'Inference in a multivariate GEV model – asymptotic properties of two test statistics'. Scand. J. of Statist. **13**.

Gomes, M.I. and Pestana, D.D. (1986). 'Non classical extreme value models'. III International Conference on Statistical Climatology, Wien, Austria.

Graça Martins, E. and Pestana, D.D. (1985). 'The extremal limit problem – extensions'. V. Panonian Symp. on Math. Statist., Visegrád, Hungary.

Hall, P. (1978). 'Representations and limit theorems for extreme value distributions'. J. Appl. Probab. **15**, 639-644.

Leadbetter, M.R., Lindgrèn, G. and Rootzèn, H. (1983). Extremes and Related Properties of Random Sequences and Processes. Springer-Verlag, New York.

Mejzler, D. (1956). 'On the problem of the limit distributions for the maximal term of a variational series'. Lvov. Politehn. Inst. Naucn. Zap. (Fiz.-Mat.) **38**, 90-109.

Mejzler, D. and Weissman, I. (1969). 'On some results of N.V. Smirnov concerning limit distributions for variational series'. Ann. Math. Statist. **40**, 480-491.

Pickands, J. III (1975). 'Statistical inference using extreme order statistics'. Ann. Statist. **3**, 119-131.

Smith, R.L. (1984). 'Threshold methods for sample extremes'. In J.Tiago de Oliveira (ed.). Statistical Extremes and Applications, D. Reidel, 621-638.

van Montfort, M.A.J. and Gomes, M.I. (1985). 'Statistical choice of extremal models for complete and censored data'. J. Hydrology **77**, 77-87.

Weissman, I. (1975). 'Multivariate extremal processes generated by independent, non-identically distributed random variables'. J. Appl. Probab. **12**, 477-487.

Weissman, I. (1978). 'Estimation of parameters and large quantiles based on the k largest observations'. J. Amer. Statist. Assoc. **73**, 812-815.

# LARGE DEVIATIONS AND BAHADUR EFFICIENCY OF SOME RANK TESTS OF INDEPENDENCE

Piotr Bajorski and Teresa Ledwina
Institute of Mathematics, Technical University of Wrocław,
Wyb. Wyspańskiego 27, 50-370 Wrocław,
Poland

ABSTRACT. Large deviations for a class of rank tests of bivariate independence against positive quadrant dependence are derived. The test statistics are closely related to a function-valued measure of dependence (so-called monotonic dependence function). Some efficiency comparisons of new tests to Spearman's rho are given under bivariate dependence models introduced recently by Lawrance and Lewis and by Raftery.

## 1. INTRODUCTION

Classical tests for testing independence as Kendall's tau and Spearman's rho e.g. are based on estimators of some measures of dependence. In Ledwina (1986) two new rank tests of independence have been introduced in this way, also. More precisely, the new tests are closely related to a new measure of dependence introduced by Kowalczyk and Pleszczyńska (1977). In this paper, we generalize the class of test statistics as well as large deviation results derived in Ledwina (1986).

To define the statistics under consideration, let us denote by $(X_1,Y_1),\dots,(X_n,Y_n)$ a set of i.i.d. continuous random vectors and let $R_i(S_i)$ be the rank of $X_i(Y_i)$ among $X_1,\dots,X_n(Y_1,\dots,Y_n)$. Moreover, let for $p\in(0,1)$



*P. Bauer et al. (eds.), Mathematical Statistics and Probability Theory, Vol. B, 11–23.*

$$T_n^o(p) = (n(n+1))^{-1} \sum_{i=1}^{n} R_i(p - I(S_i \le (n+1)p)) \tag{1.1}$$

where $I(A)$ is the indicator of the set $A$. Then the test statistic is of the form

$$W_n = \sup_{0<p<1} w(p) T_n^o(p) , \tag{1.2}$$

where the weight function $w$ satisfies the following assumptions

A. $w(p) = I(\varepsilon \le p \le 1-\varepsilon) w^*(p)$, $0 \le \varepsilon < 1/2$, where $w^*(p)$ is continuous on $(0,1)$. Moreover, if $\varepsilon = 0$ then $w^*(p)p(1-p) \to 0$ as $p(1-p) \to 0$.

In Section 3 large deviations for $W_n$ are derived under the independence hypothesis. Moreover, we shall present some exact Bahadur efficiency of Spearman's rho to $W_n$ with the following weights

$$w(p) \equiv 1 \quad \text{and} \quad w(p) = I(\varepsilon \le p \le 1-\varepsilon)\{p(1-p)\}^{-1/2}, \; 0<\varepsilon<1/2. \tag{1.3}$$

This choice of weight functions is related to availability of expansions of the large deviations rate.

To simplify the presentation and to have consistent notation with Ledwina (1986) we shall denote by $t_n$ and $t_n^\varepsilon$ the statistic $W_n$ with the first and the second mentioned weight, respectively. Generally, $t_n^\varepsilon$ is more efficient than $t_n$. In Section 4 some numerical values of the efficiencies are provided for some dependence models introduced by Lawrance and Lewis (1983) as well as by Raftery (1984). For both distributions $t_n^\varepsilon$ is more efficient than rho. Under Raftery's distribution $t_n$ dominates rho, also. Under the second distribution the situation is reversed. Note also that in Ledwina (1986) two statistics of the form (1.2) with weights (1.3) and with $T_n^o(p)$ replaced by

$$T_n(p) = n^{-2} \sum_{i=1}^{n} R_i(p - I(S_i \le np)) \tag{1.4}$$

have been considered. A motivation to replace $T_n(p)$ by $T_n^o(p)$ is $T_n^o(p) \le 1-p$ for all $p \in (0,1)$ and consequently for larger class of weight functions large deviations of $W_n$ exists. Anyway, under (1.3) this modification is not essential. Hence, some further examples of efficiencies of $W_n$ to rho can be found in Ledwina (1986).

Note also that in Ledwina (1986a) it has been shown that the local (as the alternative approaches the null hypothesis) exact Bahadur efficiency of $t_n$ to rho and the limiting (as the significance level approaches 0) Pitman efficiency coincide. Moreover, the local optimality in the sense of Bahadur of $t_n$ and $t_n^\varepsilon$ is studied in Bajorski (1986).

## 2. PRELIMINARIES

First we state some simple but useful properties of $T_n^o(p)$.

$$T_n^o(0) = T_n^o(1) = 0 ,$$
$$T_n^o(p) \le \min\{p/2, (1-p)/2\} , \tag{2.1}$$

$$T_n(p) - 1/n \le (1+1/n) T_n^o(p) \le T_n^o(p) . \tag{2.2}$$

Now we shall describe large deviations for $w(p)T_n^o(p)$ . To formulate the result the following notation are introduced. For two df's on the plane, say $H_1$, $H_2$ , the Kullback-Leibler information number of $H_1$ with respect to $H_2$ is denoted by $K(H_1,H_2)$ . For a set $E$ of df's define

$$K(E,H_2) = \inf\{K(H_1,H_2) : H_1 \in E\} .$$

Let $B$ be the set of all bivariate continuous df's having uniform marginals on $(0,1)$ . $H_o \in B$ is defined by $H_o(u,v) = uv$ , $(u,v) \in (0,1)^2$. Put $a_p(u,v) = u(p-I(v \leq p))$ and define, for $H \in B$

$$T(H,p) = \iint a_p dH \quad , \quad \bar{t}(p) = \sup\{T(H,p) : H \in B\} \quad ,$$

$$A_t(p) = \{H \in B : T(H,p) \geq t\} \quad , \quad I(t,p) = K(A_t(p),H_o) \ .$$

Throughout the paper when the range of integration is unspecified it is understood to be $(0,1)$ . Moreover, since $2T(H,p)/p(1-p)$ equals to the measure of dependence introduced by Kowalczyk and Pleszczyńska (1977), we have $|T(H,p)| \leq p(1-p)/2$ for all $H \in B$ .

From Theorem 1 of Woodworth (1970) one can easily derive large deviations for $w(p)T_n(p)$ (cf. Ledwina, 1986, Sec. 4.2) . Moreover, by (2.2), it is also seen that large deviations of $w(p)T_n(p)$ and $w(p)T_n^o(p)$ coincide. Hence

LEMMA 2.1. Under $H_o$ , for every fixed $p \in (0,1)$ and every sequence $\{r_n\}$ of real numbers such that $r_n \to 0$ as $n \to \infty$ it holds

$$-\lim_{n\to\infty} n^{-1} \log P(w(p)T_n^o(p)+r_n \geq t) = I(t/w(p),p) \tag{2.3}$$

for $t \in (0,w(p)\bar{t}(p))$ .

Next lemma we state will be used to describe large deviations of $W_n$ in terms of the $I(t,p)$ function. Due to (2.1) and the continuity of $w^*$ , the result follows by some elementary argument and therefore its proof is omitted. To formulate the lemma define

$$A_t^o(p) = \{H \in B : w(p)T(H,p) \geq t\} \ ,$$

$$A_t = \{H \in B : \sup_{0<p<1} w(p)T(H,p) \geq t\} \ ,$$

$$\bar{t} = \sup_{0<p<1} \{w(p)p(1-p)/2\} \ , \ M = \{p : \frac{w(p)}{2} \min(p,(1-p)) \geq t\} \ .$$

LEMMA 2.2. For $t \in (0,\bar{t})$ it holds

$$K(A_t, H_o) = \inf_{0<p<1} K(A_t^o(p), H_o) = \inf_{p\in M} I(t/w(p), p) .$$

We close this section with the following technical lemma.

LEMMA 2.3. For arbitrary $z$'s converging to $0$ from the right

$$\lim_{z\to 0} \inf_{p\in M} I(t/w(p) - z, p) = \inf_{p\in M} I(t/w(p), p) .$$

<u>Proof</u>. Since $1/w(p)$ is bounded on $M$, by Lemma 2.2 we have

$$K(A_{t-z'}, H_o) \le \inf_{p\in M} I(t/w(p)-z, p) \le K(A_{t-z''}, H_o)$$

where $z' = z \sup\{w(p) : p\in M\}$, $z'' = z \inf\{w(p) : p\in M\}$. So, the proof will be concluded if we establish continuity from the left of the function $t \to K(A_t, H_o)$. To justify the continuity we extend the definition of $T(H,p)$, $H\in B$, on the set of all bivariate df's. For arbitrary $H(x,y)$ with marginals $F(x)$ and $G(y)$ put

$$T^e(H,p) = \int_{-\infty}^{\infty}\int_{-\infty}^{\infty} F(x)(p-I(G(y)\le p))dH(x,y) .$$

It is known that $\sup_{0<p<1} T^e(H,p)$ is continuous in the uniform topology (cf. Ledwina, 1986). If $F$ and $G$ are continuous df's then $T^e(H,p) = T(\bar{H},p)$, where $\bar{H}(u,v) = H(F^{-1}(u), G^{-1}(v))$, $(u,v) \in (0,1)^2$. Obviously $\bar{H}\in B$. Moreover, in this case $|T^e(H,p)| \le \bar{t}(p) = p(1-p)/2$.

Let $H_n \ll H_o^* = F\times G$, $H \ll H_o^*$ and suppose $H_n$ converges to $H$ uniformly. By the above

$$|T^e(H_n,p) - T^e(H,p)| \le p(1-p) .$$

Hence one easily gets

$$\sup_{0<p<1} w(p)|T^e(H_n,p) - T^e(H,p)| \to 0 .$$

So, $\sup_{0<p<1} w(p)T^e(H,p)$ is continuous on the set $\Gamma = \{H : K(H,H_o^*)<\infty\}$ .
Define

$$A_t^e = \{H \ll H_o^* : \sup_{0<p<1} w(p)T^e(H,p) \geq t\} .$$

By Lemma 3.3 of Groeneboom et al. (1979) the function $t \to K(A_t^e,H_o^*)$ is continuous from the left. Since $K(A_t^e,H_o^*) = K(A_t,H_o)$, the proof is completed.

## 3. LARGE DEVIATIONS FOR $W_n$

THEOREM 3.1. Let $w$ satisfies the assumption A . Under $H_o$ , for $t\in(0,\bar{t})$ we have

$$\lim_{n\to\infty} n^{-1}\log P(W_n \geq t) = -K(A_t,H_o) .$$

Proof. Similarly as in Ledwina (1986) our proof is based on standard reasoning (cf. Bahadur (1971), e.g.). By Lemma 2.1 ,

$$\liminf_{n\to\infty} n^{-1}\log P(W_n\geq t) \geq - \inf_{p\in M} I(t/w(p),p) .$$

Turn now to the orher inequality. For each integer $k$ take a partition $0=p_o<p_1<\ldots<p_k<p_{k+1}=1$ of the interval $[0,1]$ such that $\max\{(p_{i+1}-p_i), 0\leq i\leq k\} \to 0$ as $k \to \infty$ . Then for sufficiently large $k$ we get

$$P(W_n\geq t) \leq k \max_{1\leq i\leq k-1} P(T_n(p_i) \geq t/w(p_i) - a(k)) ,$$

where $a(k) = \max\{ t[1/w(p_i)-1/w_i] + (p_{i+1}-p_i)/2 , 1\le i\le k-1\}$ while $w_i = \sup\{w(p) : p_i\le p\le p_{i+1}\}$ . By Lemma 2.1 it holds

$$\limsup_{n\to\infty} n^{-1}\log P(W_n\ge t) \le - \inf_{p\in M} I(t/w(p)-a(k),p) .$$

Since $a(k)>0$ and the uniform continuity of $1/w(p)$ on $M$ implies $a(k) \to 0$ as $k \to \infty$ , the proof is concluded by application of Lemmas 2.3 and 2.2 .

## 4. EXAMPLES

In this section we shall present some exact Bahadur efficiencies of $t_n$ and $t_n^\varepsilon$ , i.e. $W_n$ with weights (1.3), to Spearman's rho under some dependence models introduced by Lawrance and Lewis (1983) as well as by Raftery (1984). We restrict ourself to state some properties and to present some crucial formulae, only. All necessary calculations are contained in Bajorski and Ledwina (1986), Section 5.

Bahadur efficiencies of $t_n$ and $t_n^\varepsilon$ to Spearman's rho and Kendall's tau under some dependence models have been previously derived and discussed in Ledwina (1986). Since rho and tau are locally equivalent under both alternatives considered in this section, we compare in this paper $W_n$ and rho, only.

Before a presentation of efficiencies for the two alternatives entioned at the beging, we recall some necessary facts and notation.

Let $f(t,S)$ , $f(t,T)$ and $f(t,T^\varepsilon)$ stand for indices of large deviations for Spearman's rho, $t_n$ and $t_n^\varepsilon$ , respectively. By Kremer (1979) and Ledwina (1986), it holds

$$f(t,S) = .5t^2 + o(t^2) ,$$

$$f(t,T) = 24t^2 + o(t^2) ,$$

$$f(t,T^{\varepsilon}) = 6t^2 + o(t^2) ,$$

as $t \to 0$ .

Moreover, let $b(H,S)$ , $b(H,T)$ , $b(H,T^{\varepsilon})$ denote the strong limits under the alternative $H$ of the considered statistics. By Woodworth (1970) and Ledwina (1986) we have

$$b(H,S) = 12 \iint uvd\bar{H}(u,v) - 3 ,$$

$$b(H,T) = \sup_{0<p<1} \iint u(p-I(v\le p))d\bar{H}(u,v) ,$$

$$b(H,T^{\varepsilon}) = \sup_{\varepsilon\le p\le 1-\varepsilon} [p(1-p)]^{-1/2} \iint u(p-I(v\le p))d\bar{H}(u,v) ,$$

where $\bar{H}(u,v) = H(F^{-1}(u),G^{-1}(v))$ , $F$, $G$ are marginals of $H$ while $F^{-1}$, $G^{-1}$ their inverses.

By the well known Bahadur result (cf. Bahadur, 1971, Theorem 7.2, e.g.) Bahadur slope of any of the above statistics for testing against the alternative $H$ is $2f(b(H,.),.)$ . Besides, by definition, the exact Bahadur efficiency of two statistics is the ratio of their Bahadur slopes. From the above formulae it is immediatly seen that $f(b(H,T),T)/f(b(H,T^{\varepsilon}),T^{\varepsilon}) \to r \le 1$ , provided $H$ approaches to the null distribution. It simply means that $T^{\varepsilon}$ is always locally at least as good as $T$ .

## 4.1. Bahadur efficiencies under the Lawrance-Lewis alternative

The distribution introduced by Lawrance and Lewis (1983), p. 1186, has the density

$$h_z(u,v) = \frac{(uv)^{(1-z)/z}}{z} + \frac{(uv)^{-z/(1+z)}}{1+z} I(v^{1/z}\le u\le v^{z}) , \qquad (4.1)$$

where $0 \leq u,v \leq 1$ , $0 \leq z \leq 1$ . By straightforward calculations it is seen that for all $z$ (4.1) shows positive quadrant dependence.

Since (4.1) has uniform marginals, the correlation coefficient for it equals to $b(H_z,S)$ , where $H_z$ is the corresponding d.f. To simplify notation we shall replace $b(H_z,.)$ by $b(z,.)$ in what follows. By Lawrance and Lewis (1983), p. 1186,

$$b(z,S) = \frac{3z(1-z)(8+7z+z^2)}{(1+z)^2(2+z)^2} .$$

Moreover, by Lemma 3 of Lehmann (1966) independence is equivalent to $z=0$ or $z=1$ .

Let

$$A(z,p) = \frac{zp^{2/z}}{4+2z} + \frac{p}{2} - \frac{zp^{1/z}}{1+z} - \frac{p^{(1+z)}}{(2+z)(1+z)} .$$

A little algebra shows that

$$b(z,T^{\varepsilon}) = \sup_{\varepsilon \leq p \leq 1-\varepsilon} A(z,p)[p(1-p)]^{-1/2} ,$$

$$b(z,T) = \sup_{0<p<1} A(z,p) = \frac{zp_o-(1-z)p_o^{2/z}}{2+2z} + \frac{(1-z-z^2)p_o^{1/z}}{(1+z)^2} ,$$

where $p_o$ stands for the sole root from $(0,1)$ of the equation

$$\frac{p^{(1/z\ -1)}}{1+z} + \frac{p^z-p^{(2/z\ -1)}}{2+z} = \frac{1}{2} .$$

Table 1 contains some numerical values under $\varepsilon = 0.001$ , where $e(z,.,S)$ denotes exact Bahadur efficiency of . to S .

Table 1. Bahadur efficiencies under the Lawrance-Lewis alternative

| $z$ | $b(z,S)$ | $e(z,T,S)$ | $e(z,T^{\varepsilon},S)$ |
|---|---|---|---|
| .001 | .0060 | .798 | 69.2 |
| .002 | .0119 | .770 | 34.3 |
| .003 | .0178 | .760 | 22.6 |
| .004 | .0237 | .756 | 16.9 |
| .007 | .0411 | .749 | 9.7 |
| .011 | .0638 | .745 | 6.3 |
| .017 | .0967 | .743 | 4.2 |
| .029 | .1590 | .741 | 2.6 |
| .880 | .1614 | .856 | 1.13 |
| .930 | .0939 | .860 | 1.18 |
| .965 | .0468 | .862 | 1.21 |
| .995 | .0067 | .864 | 1.23 |

## 4.2. Bahadur efficiencies under the Raftery alternative

The uniform representation, i.e. the d.f. $\bar{H}$, of the pair of variables introduced by Raftery (1984) (cf. his formula 2.3 and Appendix) has the density

$$h(u,v) = \begin{cases} (1-v)^z(1+z(1-u)^{-1-z}) & \text{for} \quad v \geq u \ , \\ (1-v)^z & \text{otherwise} \ , \end{cases} \tag{4.2}$$

where $z \geq 0$ . Our $z$ equals to $\pi_2/(1-\pi_2)$ in Raftery's notation and for the sake of brevity of formulae we derive we consider $z \in (0,1)$ , only .

Since (4.2) defines also positively quadrant dependent variables and the correlation coefficient for (4.2) is

$$b(z,S) = z/(2+z) \ ,$$

the independence is equivalent to $z=0$ . By some straightforward calculations we get

$$b(z,T) = \frac{z}{2(1+z)} q(1-q) ,$$

where $q$ stands for the sole root from $(0,1)$ of the equation

$$(1+z)q^z - 2qz = 1-z ,$$

while

$$b(z,T^\varepsilon) = [(1-p_1)/2p_1]^{1/2}\{1+[z(1-p_1)-(1-p_1)^z]/(1-z)\} ,$$

where $p_1 = \min\{1-\varepsilon, p_2(z)\}$ and $p_2(z)$ stands for the sole root from $(0,1)$ of the equation

$$(1-p)^z(1+2pz) - p(1-2p)z = 1$$

If $p_2 \le 1-\varepsilon$ we get the simpler formula

$$b(z,T^\varepsilon) = p_2 z[p_2(1-p_2)]^{1/2}/(1+2p_2 z)$$

Table 2 contains some numerical values of efficiencies under $\varepsilon = 0.001$ and under the same simplified notation as in the previous example.

Table 2. Bahadur efficiencies under the Raftery alternative

| $z$ | $b(z,S)$ | $e(z,T,S)$ | $e(z,T^\varepsilon,S)$ |
|---|---|---|---|
| .001 | .0005 | 1.258 | 2.698 |
| .052 | .0252 | 1.246 | 2.560 |
| .106 | .0504 | 1.235 | 2.433 |
| .176 | .0811 | 1.221 | 2.294 |
| .254 | .1128 | 1.208 | 2.165 |
| .387 | .1621 | 1.187 | 1.991 |

Moreover, for the distribution (4.2) it can be shown that

$$\lim_{z\to 0} e(z,T,S) = 48d_o^2(1-d_o)^2 \approx 1.258$$

and

$$\lim_{z\to 0} e(z,T^{\varepsilon},S) = 48d_1^3(1-d_1) \approx 2.701 \quad ,$$

for $\varepsilon$ sufficiently small ($\varepsilon < 1-d_1$), where $d_o$ and $d_1$ are sole roots from $(0,1)$ of the equations $2d-\ln d = 2$ and $2d^2+d+\ln(1-d) = 0$, respectively.

ACKNOWLEDGEMENTS. We are grateful to A.J. Lawrance for sending us the reprint of Lawrance and Lewis (1983).

REFERENCES

Bahadur,R.R. (1971). Some Limit Theorems in Statistics. SIAM, Philadelphia.

Bajorski,P. (1986), 'Local Bahadur optimality of some rank tests of independence'. Tech.Report No. 28 , Institute of Mathematics, Technical University of Wrocław (to appear: Statistics and Probability Letters 5, 1987).

Bajorski,P. and Ledwina,T. (1986). 'Large deviatons and Bahadur efficiency of some rank tests of independence'. Tech.Raport No. 37 , Institute of Mathematics, Technical University of Wrocław.

Groeneboom,P., Oosterhoff,J. and Ruymgaart,F.H. (1979). 'Large deviation theorems for empirical probability measures'. Ann. Probability 7 , 553-586.

Kowalczyk,T. and Pleszczyńska,E. (1977). 'Monotonic dependence functions of bivariate distributions'. Ann. Statist. 5 , 1221-1227.

Kremer,E. (1979), 'Approximate and local Bahadur efficiency of linear rank tests in the two-sample problem'. Ann. Statist. 7, 1246-1255.

Lawrance,A.J. and Lewis,P.A.W. (1983). 'Simple dependent pairs of exponential and uniform random variables'. Operations Research 31, 1179-1197.

Ledwina,T. (1986a). 'On the limiting Pitman efficiency of some rank tests of independence'. J. Multivar. Anal. 20 ,265-271.

Ledwina,T. (1986). 'Large deviations and Bahadur slopes of some rank tests of independence'. Sankhyā, Ser. A, 48, 188-207.

Lehmann,E. (1966). 'Some concepts of dependence'. Ann. Math. Statist. 37, 1137-1153.

Raftery,A.E. (1984). 'A continuous multivariate exponential distribution'. Commun. Statist. Theor. Meth. 13, 947-965.

Woodworth,G.G.(1970). 'Large deviations and Bahadur efficiency of linear rank statistics'. Ann. Math. Statist. 41, 251-283.

# ON SOME OBSERVING AND REPLACEMENT POLICIES

Dragan M. Banjevic
University of Belgrade
Faculty for Natural and
Mathematical Sciences
Studentski Trg 13
11000 Beograd, Yugoslavia

Ranko R. Nedeljkovic
University of Belgrade
Faculty for Traffic and
Transport Engineering
Vojvode Stepe 305
11000 Beograd, Yugoslavia

ABSTRACT. The interval between two failures of an equipment is a random variable. In order to keep the equipment in operation, the preventive replacements are performed. In this paper it is considered that there are certain periods of time in which the equipment works unobserved. There are losses due to failures, costs of the preventive replacements, as well as of the control of the equipment. The loss is also caused when the equipment fails during an unobserved period. The bounds on the loss function and optimal observing and replacement policies are discussed.

## 1. INTRODUCTION

During the operation of an equipment, the failures occur. In order to minimize losses caused by them, the preventive replacements are performed. We shall consider that they are performed momentarily and that afterwards the equipment operates as a new one. The life lengths of the equipment are nonnegative, independent, identically distributed random variables. In this paper a model, in which during certain period of time the equipment works unobserved, is discussed.

A model close to this one, but without periods of non-observation, is presented in Barlow and Proschan (1965).

Here, a model close to real situations is introduced: the operation brings the profit so it is important to keep the equipment in work, and, on the other hand, it might be expensive or technically difficult to control its work all the time.

The equipment starts operating at the moment $t = 0$, but, during the interval of time $[0,R)$, it works unobserved. If in this period the failure occurs, the equipment does not work till the moment $t = R$, when the failure is registered and the replacement performed. In the interval $[R,T)$ the equipment is under observation, so that every

---

AMS 1980 subject classification. Primary 62N05, secondary 60K10, 62C20
Key words and phrases. Loss function, optimal strategy, optimal minimax strategy, replacement policies, periods of non-observation.



*P. Bauer et al. (eds.), Mathematical Statistics and Probability Theory, Vol. B, 25–31.*

failure is immediately noticed and the replacement done. If, however, the failure does not occur up to the moment $t = T$, the preventive replacement is performed.

Let $C(t)$ be the expected total cost of the control and replacements on the interval $[0,t]$. Since we want to observe the equipment during a long period of time, it is natural to take $\lim_{t\to\infty} C(t)/t$ as an optimization criterion, i.e. it is preferable to optimize the total cost of the control and replacements, per unit of time.

We also consider a model in which the observing period is immediately after the replacement, as well as the model with two periods of control.

## 2. THE LOSS FUNCTION

We shall call $L(R,T) = \lim_{t\to\infty} C(t)/t$ the loss function. Here, $R$ is the length of the unobserving interval, $T$ is the moment of the preventive replacement if the failure does not occur before it, $0 \leq R \leq T \leq \infty$. A pair $(R,T)$ will be called the strategy. Our main goal is to obtain optimal values $R^o, T^o$ so that $L(R^o,T^o) = \min_{R,T} L(R,T)$.

It can be shown that $L(R,T) = EV/EY$, where $V$ is the cost between two replacements and $Y$ is the length of time between two replacements.

Let $a,b,c,d,e$ be as follows:

a - the cost of nonoperating of the equipment, per unit of time;
b - the cost of the replacement caused by a failure in an unobserved period;
c - the cost of the observing, per unit of time;
d - the cost of the replacement caused by a failure in an observed period;
e - the cost of the preventive replacement.

We shall take $b \geq d \geq e > 0$ as a natural assumption.

Theorem 1.

$$L(R,T) = EV/EY = c + \frac{-a\int_0^R (1-F(x))dx + (b-d)F(R)}{\int_R^T (1-F(x))dx + R} + \frac{(d-e)F(T) + (a-c)R + e}{\int_R^T (1-F(x))dx + R} \quad (1)$$

where $F$ is the life distribution.

Proof.

Let $X$ be the life length, a nonnegative random variable with distribution $F$. Then $V$ and $Y$, introduced above, are:

$$V = [a(R-X)+b]I\{X<R\} + [c(X-R)+d]I\{R \leq X<T\} + [c(T-R)+e]I\{X \geq T\};$$
$$Y = RI\{X<R\} + XI\{R \leq X<T\} + TI\{X \geq T\}.$$

Using that

$$\int_0^T (1-F(x))dx = \int_0^T x\,dF(x) + T(1-F(T)),$$

we have

$$EV = c\int_0^T (1-F(x))dx - (a+c)\int_0^R (1-F(x))dx + (b-d)F(R) +$$
$$+ (d-e)F(T) + aR + e ,$$
$$EY = RF(R) + \int_R^T x dF(x) + T(1-F(T)) = \int_R^T (1-F(x))dx + R$$

which gives the result.

Some real situations point out special cases such as:

(a) Model with no unobserved period. The loss function is $L(0,T)$.

(b) Model without observing periods. At the moment $t = R$ the replacement upon failure or the preventive replacement is performed. The loss function is $L(R,R)$.

(c) Model with no preventive replacements, i.e. $T = \infty$, and the loss function is $L(R,\infty)$.

Each of these special cases requires separate discussions. Here we shall discuss some properties of the function $L(R,T)$ in general. As an important example, the exponential distribution for the life length will be used.

## 3. THE BOUNDS ON THE LOSS FUNCTION. SOME OPTIMAL STRATEGIES.

Let us start our analysis with the bounds on the function $L(R,T)$. We shall assume that $m = EX$ is finite.

Using $b \geq d \geq e$, we obtain:

$$L(R,T) \leq a + b/R \tag{2}$$

Since $m = EX = \int_0^\infty (1-F(x))dx$ implies that

$\int_0^T (1-F(x))dx \leq \min\{T,m\}$, we have:

$$L(R,T) \geq \frac{a(R-\min\{m,R\}) + e}{\min\{T,m+R\}} \geq \frac{a(R-m) + e}{m + R} = a + \frac{e-2am}{m+R} \tag{3}$$

Theorem 2. Let $e \geq 2am$. Then

$$\min_{R,T} L(R,T) = L(\infty,\infty) = a ,$$

i.e. $(\infty,\infty)$ is the optimal strategy.

Proof.

It is easy to see that $L(\infty,\infty) = \lim L(R,T) = a$. For $e \geq 2am$, from (3), we have $L(R,T) \geq a$, i.e. $L(R,T) \geq L(\infty,\infty)$.

In the case $e < 2am$, "finite" strategies are possible.

Theorem 3. Let $e < 2am$. Then:

$$R^o \geq \frac{e}{\min\{a, c + d/m\}} - m , \quad T^o \geq \frac{e}{\min\{a, c + d/m\}}$$

Proof.

It is obvious that $L(0,\infty) = c + d/m$, and, from (3),

$$\min_T L(R,T) \geqslant e/(m+R), \quad \min_R L(R,T) \geqslant e/T. \quad \text{But,}$$

$$\min_{R,T} L(R,T) \leqslant \min\{L(0,\infty),L(\infty,\infty)\} = \min\{c + d/m,a\}. \text{ Now}$$

$e/(m+R^o) \leqslant \min\{c + d/m,a\}$, $e/T^o \leqslant \min\{c + d/m,a\}$ give the result.

## 4. APPLICATION TO THE EXPONENTIAL LIFE DISTRIBUTION

Let $F(x) = 1 - \exp(-x/m), \quad x \geqslant 0.$

Theorem 4. The optimal strategy is $(\infty,\infty)$ if and only if

$$a \leqslant \min\{b/m, c + d/m\}.$$

Proof.

The strategy $(\infty,\infty)$ is the best one if and only if $L(R,T) \geqslant L(\infty,\infty) = a$, for all $R$ and $T$, which is equivalent to the inequality:

$$(c-a)\int_0^T(1-F(x))dx - c\int_0^R(1-F(x))dx + (b-d)F(R) + \\ + (d-e)F(T) + e \geqslant 0. \tag{4}$$

For the exponential life distribution (4) becomes:

$$(c + d/m - a)F(T) + (e/m)(1-F(T)) + (b/m-(c+d/m))F(R) \geqslant 0 \tag{5}$$

From (5), the strategy $(\infty,\infty)$ is better than the strategy $(0,\infty)$ if and only if $c + d/m \geqslant a$. If $b/m < a$ we have, from (5), that the strategy $(R,T)$ where $R$ and $T$ are finite and sufficiently large is better than $(\infty,\infty)$. So, $c + d/m \geqslant a$ and $b/m \geqslant a$ are necessary conditions for $(\infty,\infty)$ to be the best strategy. But these conditions are sufficient, too, because, for all $R$ and $T$:

$$(c + d/m-a)F(T) + (e/m)(1-F(T)) + (b/m - (c+d/m))F(R) \geqslant \\ \geqslant (c + d/m - a)(F(R)) + (e/m)(1-F(T)) + (b/m - (c + d/m))F(R) = \\ = (b/m - a)F(R) + (e/m)(1-F(T)) \geqslant 0.$$

Theorem 5. The optimal strategy is $(0,\infty)$ if and only if

$$c + d/m \leqslant \min\{a,b/m\}.$$

Proof.

The strategy $(0,\infty)$ is the best one if and only if $L(R,T) \geqslant L(0,\infty) = c + d/m$, for all $R$ and $T$, which is equivalent to the inequality:

$$(b-d)F(R) + (d-e)F(T) + (a - (c + d/m))R + e \geqslant$$
$$\geqslant (a - d/m)\int_0^R (1-F(x))dx + (d/m)\int_0^T (1-F(x))dx. \quad (6)$$

For the exponential life distribution (6) becomes:

$$(b/m - a)F(R)+(c+d/m-a)\ln(1-F(R))+(e/m)(1-F(T)) \geqslant 0. \quad (7)$$

The strategy $(0,\infty)$ is better than the strategy $(R,\infty)$ if we take $T = \infty$ in (7), which is equivalent to:

$$(b/m - a) + (c + d/m - a)(\ln(1-F(R))/F(R)) > 0. \quad (8)$$

But, $\ln(1-F(R))/F(R) < -1$ for all $R$, and $\ln(1-F(R))/F(R) \to -\infty$ when $R \to +\infty$, which implies that (8) holds if and only if $b/m \geqslant c + d/m$ and $c + d/m \leqslant a$, i.e. $c + d/m \leqslant \min\{b/m, c+d/m\}$. It means that $c + d/m \leqslant \min\{a,b/m\}$ is necessary for $(0,\infty)$ to be the best strategy. But this condition is also sufficient for (7) to hold.

According to the theorems 4 and 5, if $a > \min\{b/m, c+d/m\}$ and $c + d/m > \min\{a,b/m\}$, which is equivalent to:

$$b/m < \min\{a, c+d/m\}, \quad (9)$$

then one of the strategies $(0,T)$, $(R,\infty)$, or $(R,T)$ for $0<R\leqslant T<\infty$, is the best one.

A short analysis shows that the sign of $\partial L(R,T)/\partial T$ depends only on $R$, which implies that $L(R,T)$ is a monotone function in $T$, and that means that one of the strategies $(R,R)$ or $(R,\infty)$ is the optimal one, i.e.

$$\min_{R,T} L(R,T) = \min\{\min_R L(R,R), \min_R L(R,\infty)\}.$$

On condition given in (9) it can be shown that:

$$\min_R L(R,R) = L(R_1^o,R_1^o) = a + \frac{(b-e-am)F(R_1^o) + e}{R_1^o}$$

where $R_1^o$ is the solution of the equation:

$$(1 + R/m)\exp(-R/m) = (b-am)/(b-e-am). \quad (10)$$

Also, it can be shown that

$$\min_R L(R,\infty) = L(R_2^o,\infty) = c + \frac{(a-c)R_2^o+(b-d-am)F(R_2^o)+d}{m(1-F(R_2^o)) + R_2^o}$$

where $R_2^o$ is the solution of the equation:

$$(1+R/m+(b-am)/(b-d-cm))\exp(-R/m) = (b-am)/(b-d-cm) \quad (11)$$

Finally, $\min_{R,T} L(R,T) = \min\{L(R_1^o,R_1^o), L(R_2^o,\infty)\}$, i.e. one of the strategies $(R_1^o, R_1^o)$, $(R_2^o,\infty)$ is the best one.

Example. Let $F(x) = 1 - \exp(-x)$, $x \geq 0$, $(m=1)$, and $a=15$, $b=12$, $c=10$, $d=8$, $e=6$. Find the optimal strategy.
The condition in (9) is fulfilled. From equation (10) we obtain $R_1^o = 2.28927$ and $L(R_1^o, R_1^o) = 13.7338$. Also, from (11), we get $R_2^o = 1.92392$ that gives $L(R_2^o, \infty) = 13.9740$. Now, $\min L(R,T) = 13.7338$ which implies that $(2.28927, 2.28927)$ is the optimal strategy.

## 5. THE OPTIMAL MINIMAX STRATEGY.

The basic principle of this method is well known: choose the worst possible situation and try to optimize it. In our case, it would be considered that we do not know the life distribution, except, perhaps, its mean. We shall maximize the loss function $L(R,T)$ and, then, find optimal values $R$ and $T$ so that they minimize losses, i.e. we are trying to find

$$\min_{R,T} \max_{F \in \mathcal{F}_m} L(R,T)$$

where $\mathcal{F}_m$ is a class of all possible distributions with the mean $m$.

Theorem 6. The optimal minimax strategy for the class $\mathcal{F}_m$ of the life distributions is: (a) $(0,\infty)$, if $c+d/m < a$
(b) $(\infty,\infty)$, if $c+d/m \geq a$.

Proof.
Let $\overline{L}(R,T) = \sup_{F_m} L(R,T,F_m)$, $F_m \in \mathcal{F}_m$. $(\overline{L}(0,\infty) = c+d/m$, $\overline{L}(\infty,\infty) = a)$

Then

$$\inf_{0 \leq R \leq T \leq \infty} L(R,T) = \min\{\overline{L}(0,\infty), \overline{L}(\infty,\infty), \inf_{0 \leq R \leq T < \infty} \overline{L}(R,T)\}.$$

It will be shown that

$$\overline{L}(R,T) \geq a. \tag{12}$$

For fixed $F_m$ the inequality $L(R,T,F_m) \geq a$ becomes:

$$(c-a)\int_0^T (1-F_m(x))dx - c\int_0^R (1-F_m(x))dx + (b-d)F(R) + (d-e)F(T) + e \geq 0 \tag{13}$$

Let

$$F^{\alpha}(x) = 1 - ((\alpha x+1)^{1+1/\alpha m})^{-1}, \quad x \geq 0, \ \alpha > 0.$$

It is easy to see that $EX = m$, and that, for every $x>0$, $F^{\alpha}(x) \to 1$, when $\alpha \to \infty$, and that, for every finite $A$,

$$\int_0^A (1-F^{\alpha}(x))dx \to 0, \text{ when } \alpha \to \infty.$$

Using the properties of $F^{\alpha}(x)$ we can see that, for any fixed R and T, the inequality (13) is fulfilled for sufficiently large $\alpha$. That implies (12) and

$$\inf_{0 \leq R \leq T \leq \infty} \overline{L}(R,T) = \min\{\overline{L}(0,\infty), \overline{L}(\infty,\infty)\} = \min\{c+d/m, a\}.$$

REFERENCE

Barlow R. E. and Proschan F.(1965) Mathematical theory of reliability, Wiley, New York.

ON STRONG CONSISTENCY OF KERNEL ESTIMATORS UNDER DEPENDENCE ASSUMPTIONS

J. Deddens, M. Peligrad*, T. Yang**
Department of Mathematical Sciences
University of Cincinnati
Cincinnati, Ohio 45221-0025

ABSTRACT. Some considerations on the uniform consistency of the kernel estimator of a density and of a regression function are made for certain dependent samples.

## 1. INTRODUCTION

Let $X_1, X_2, \ldots, X_n$ be identically distributed p-dimensional random vectors with unknown density function $f(x) = f(x_1, \ldots, x_p)$. We consider estimators $f_n(x)$ of the density $f(x)$ of the following form (Rosenblatt (1956), Parzen (1962))

$$(1.1) \qquad f_n(x) = \frac{1}{nh_n^p} \sum_{j=1}^{n} K\left(\frac{x-X_j}{h_n}\right)$$

where $h_n \to 0$ as $n \to \infty$ and $K(y)$ is a real-valued Borel measurable function on $R^p$.

Let $(X_i, Y_i)$, $1 \leq i \leq n$, be $R^p \times R$ valued random vectors. Assume there is a function r, $r(X_i) = E(Y_i | X_i)$. Define the estimators $r_n(x)$ of $r(x)$ by (Nadaraja (1964), Watson (1964))

$$(1.2) \qquad r_n(x) = \frac{\sum_{i=1}^{n} Y_i \, K\left(\frac{x-X_i}{h_n}\right)}{\sum_{i=1}^{n} K\left(\frac{x-X_i}{h_n}\right)}$$

*partially supported by a NSF grant.
**partially supported by a Taft grant.



*P. Bauer et al. (eds.), Mathematical Statistics and Probability Theory, Vol. B, 33–41.*

where $h_n \to 0$ as $n \to \infty$ and $K(y)$ is a real valued Borel measurable function on $R^p$.

When sampling independently, the uniform consistency results such as: $\sup_{x \in R^p} |f_n(x)-f(x)| \to 0$ a.s. or $\sup_{x \in R^p} |r_n(x)-r(x)| \to 0$ a.s. were obtained under certain restrictions imposed on K and f and under the restriction $\frac{n h_n^p}{\text{Log } n} \to \infty$ as $n \to \infty$. These results can be found in works by Deheuvels (1974), Devroye and Wagner (1976), Silverman (1978), and Collomb (1979).

The methods for establishing these results are either direct computations or an approximation of the empirical distribution function by a Brownian Bridge, (Komlos, Major, Tusnady (1975)). All these proofs are based on exponential bounds for the tail probabilities of partial sums. The strong consistency results remain true when sampling from a stochastic process if, for instance, good exponential bounds for partial sums can be obtained. For many dependent situations such bounds are already known. The $\phi$-mixing case is treated in papers by Bosq (1975), Samur (1984), Collomb (1984), the $\phi$ and $\rho$-mixing case in Peligrad (1985), the strong mixing situation in Delecroix (1980), Philipp (1977), and the absolute regular situation in Yoshihara (1978).

In this paper we point out other classes of dependent random variables for which one can establish exponential bounds. As a consequence the uniform consistency of the estimators when sampling from a Markov process satisfying Doeblin's condition (see Rosenblatt (1971) page 209 for the definition of Doeblin's condition) can be obtained under the same size of the bandwidth used for the independent case, namely $\frac{n h_n^p}{\text{Log } n} \to \infty$.

The conclusion is that in the dependent case when the dependence is weak enough (in the sense that will be described in our applications) we expect the kernel estimators used in the independence case to perform properly. In order to illustrate this for a finite sample, we analyzed data from bird migration.

Section 2 contains our Bernstein type inequalities. In Section 3 we state the consistency results.

## 2. EXPONENTIAL BOUNDS

Let X be a random variable such that $EX = 0$. X is said to be <u>generalized</u> <u>Gaussian</u> with constant $\alpha > 0$ (Chow (1966)) if

$$E \exp (uX) < \exp (u^2\alpha^2/2)$$

for all $u \in R$.

By Stout (1974; Lemma 5.4.2, p. 302) and by Kahane (1985; page 82) we have the following implications a => b => c => d => e where:

a) X is generalized Gaussian with constant $\alpha$ ;

b) $P(|X|>x) \leq 2 \exp(-x^2/2\alpha^2)$ for all positive x ;

c) $E|X|^\nu \leq \nu\, 2^{\nu/2}\, \Gamma(\nu/2)\, \alpha^\nu$ for every $\nu > 0$ ;

d) $EX^{2m} \leq 2^{m+1}\, m!\, \alpha^{2m}$ for every integer $m \geq 0$ ;

e) X is generalized Gaussian with constant $2\alpha$.

By Stout (1974; Lemma 4.2.3, page 244) (see also Azuma (1967) or Serfling (1969)) we have:

<u>Lemma 1</u>. Let $(Z_n)_{n\geq 1}$ be a martingale adapted to the increasing $\sigma$-algebras $(F_n)_{n\geq 1}$ and let $X_n = Z_n - Z_{n-1}$ be the differences. Suppose there is a constant $C > 0$ such that

$$\max_{1\leq i\leq n} |X_i| \leq C \quad \text{a.s.}$$

Then $Z_n$ is generalized Gaussian with constant $\alpha = \sqrt{n}\, C$.

Let $(X_k)_{k\geq 1}$ be a sequence of random variables adapted to an increasing sequence of $\sigma$-algebras $(F_k)_{k\geq 1}$. Let $S_n = \sum_{i=1}^{n} X_i$. Assume there is a constant $C > 0$ such that for all $1 \leq m \leq n$,

$$(2.1) \qquad |E(S_n - S_m | F_m)| \leq C \quad \text{a.s.}$$

Martingale differences sequences obviously satisfy (2.1). Another class having this property, was pointed out by Eberlein (1985) and is characterized by a simple assumption on the Doleans measure. Assume $X_k$ is integrable and let $(m,n] \times F$ be a predictable rectangle ($m \leq n$ and $F \in F_m$). The Doleans measure of $(S_n)_{n\geq 1}$ is

$$\lambda((m,n]\times F) = E(1_F(S_n - S_m)] \quad .$$

It is easy to see that (2.1) is equivalent to the existence of a constant $C > 0$ such that for all predictable rectangles

$$(2.2) \qquad |\lambda((m,n]\times F)| \leq CP(F) \ .$$

Proposition 1. Let $(X_n)_{n>1}$ be a sequence of random variables adapted to the increasing sequence $(F_n)_{n>1}$. Assume (2.1) is satisfied, $EX_i = 0$ for every i, and that there is $D > 0$ such that $\max_{1<i<n} |X_i| < D$ a.s. Then $S_n$ is generalized Gaussian with the constant $\alpha = \sqrt{n}(2C+D)$.

Proof. We can write:

$$S_n = \sum_{i=1}^{n} [E(S_n|F_i) - E(S_n|F_{i-1})]$$

where $X_0 = 0$ and $F_0 = \{0,\Omega\}$. It is easy to see that the sequence $Z_m = \sum_{i=1}^{m} [E(S_n|F_i) - E(S_n|F_{i-1})]$, $1 < m < n$, is a martingale with respect to $\{F_m\}_{m>0}$. Moreover the differences are bounded:

$$|E(S_n|F_i) - E(S_n|F_{i-1})| <$$
$$< |E(S_n|F_i)-S_i| + |E(S_n|F_{i-1})-S_{i-1}| + |X_i| <$$
$$< 2C+D .$$

By Lemma 1 it follows that $S_n = Z_n$ is generalized Gaussian with $\alpha = (2C+D)\sqrt{n}$.

Proposition 2. Assume $X_1, X_2, \ldots, X_n$ are random variables. Let $F_m$ be the $\sigma$-algebra generated by $X_1, X_2, \ldots, X_m$. For $1 < k < n$ define

$$\phi'_k = \max_{1<m<n-k} \{\sup |P(B|A)-P(B)|,\ A \in F_m ,$$
$$P(A) \neq 0 \ ,\ B \in \sigma(X_{m+k})\} .$$

Assume $\max_{1<i<n} |X_i| < C$ a.s. and $EX_i = 0$. Then for every $n > 0$

$$P(|S_n|>x) < 2 \exp(-x^2/[2(2+\sum_{i=1}^{n} \phi'_i)^2 C^2 n]) .$$

Proof. By Iosifescu and Theodorescu (1969; Lemma 1.1.8), we have

$$|E(X_{i+k}|F_k)| < C\,\phi'_i \quad \text{a.s.},$$

for every $1 \le i \le n-k$ and $1 \le k \le n-1$. Therefore

$$|E(X_n|F_m)-S_m| \le C \sum_{i=1}^{n} \phi_i'$$

The result follows now by Proposition 1.

Remark 1. The coefficients $\phi_k'$ are more general than the usual $\phi$-mixing coefficients from Ibragimov (1962). The difference is that in the definition of $\phi_k'$, B is taken in the $\sigma$-algebra generated only by $X_{m+k}$, while in the definition of $\phi_k$, B belongs to $\sigma$-algebra generated by $(X_{m+k}, X_{m+k+1}, \ldots)$.

Remark 2. If $(X_n)_n$ is a discrete time stationary Markov process satisfying Doeblin's condition, it can be shown that $\phi_n' = \phi_n = a\zeta^n$ for some $\alpha > 1$ and $0 < \zeta < 1$.

## 3. STRONG CONSISTENCY RESULTS.

In this section we state the consistency results. The proof of the first theorem follows the same lines as the proof of Devroye-Wagner's (1976) consistency theorem for i.i.d. The essential difference is that the exponential bounds for i.i.d. are replaced by Proposition 2. The theorem extends and improves a results by Rüschendorf (1977).

Theorem 1. Let $f_n(x)$ be defined by (1.1). Suppose $K(\cdot)$ is a density on $R^p$ with $\sup_x K(x) < \infty$, $\sup_x ||x||^p K(x) < \infty$ and $|K(x+y)-K(x)| \le C||y||$, $y \in R^p$ for some $C > 0$. Suppose that

$$(3.1) \qquad h_n \to 0 \quad \text{and} \quad nh_n^p/(\log n) \sum_{i=1}^{n} \phi_i' \to \infty \text{ as } n \to \infty.$$

If f is a uniformly continuous density with $\int_{R^p} ||x||^\gamma f(x)\, dx < \infty$ for some $\gamma > 0$, then $\sup_{x\in R^p} |f_n(x)-f(x)| \to 0$.

Remark 3. If $(X_i)_{i\ge 1}$ is a Markov chain satisfying Doeblin's condition, according to Remark 2, the condition on $h_n$ becomes:

(3.2) $h_n \to 0$ and $nh_n^p/\log n \to \infty$ .

Collomb (1984) proved the consistency of $r_n(x)$ defined by (1.2), and Collomb and Härdle (1986) established rates of this strong consistency results for $\phi$-mixing sequences, assuming instead of (3.1) the existence of a sequence $1 < m_n < n$ such that

(3.3) $\sup_n n\phi_{m_n}/m_n < \infty$ and $n\, h_n^p/m_n \log n \to \infty$ as $n \to \infty$.

Peligrad (1986) also proved strong consistency results for kernel estimators for $\phi$-mixing sequences assuming instead of (3.1):

(3.4) $n\, h_n^p/(\log n)^2 \exp(6 \sum_{i=1}^{n} \phi^{1/2}(2^i)) \to \infty$ as $n \to \infty$.

All this results allows as a corollary a consistency result when sampling from Markov processes satisfying Doeblin's condition under the assumption $nh_n^p/(\log n)^2 \to \infty$. Our next theorem has as a consequence a consistency result for the same class of Markov processes under the improved condition (3.2) which is the same that is used in the independent case. Moreover the mixing coefficient used in (3.1) is more general than that one used in (3.3) or (3.4).

For the next theorem we impose the following conditions.

Let D be a compact set in E where E is a borelian set in $R^p$ and denote by $\mu$ the Lebesgue measure on $R^p$, and by $\overline{D}_\varepsilon$ an $\varepsilon$-neighborhood of D. Assume $(X_i,Y_i)_{i>1}$ satisfies

(3.5). There is $\Gamma < \infty$ such that for every Borel set B in E and $i > 1$

$$P(X_i \in B) < \Gamma\, \mu(B)$$

(3.6) There is $\Gamma^* > 0$ such that for every Borel set B in $\overline{D}_\varepsilon$ and for every $i > 1$

$$P(X_i \in B) > \Gamma^*\, \mu(B)$$

(3.7) There is $M < \infty$ such that $|Y_i| < M$ for every $i > 1$.

Assume K is a kernel of $R^p$ satisfying

Condition 1. $|K(\cdot)| \leq \overline{K} < \infty$ , $\int |K(x)|dx \leq \tilde{K} < \infty$, $||x||^m K(x) \to 0$ as $||x|| \to \infty$, $\int K(x)\, dx > 0$ and K is Lipschitz of order $\gamma$.

From Collomb's (1984) proof and Proposition 2 we have:

Theorem 2. Assume K satisfies Condition 1, $(X_i, Y_i)$ satisfies (3.5), (3.6) and (3.7), $h_n$ satisfies (3.1) and in addition r is continuous on $\overline{D}_\varepsilon$. Then

$$\sup_{x \in D} |r_n(x) - r(x)| \to 0 \text{ a.s.}$$

as $n \to \infty$.

## REFERENCES

Azuma, K. (1967). Weighted sums of certain dependent random variables. Tohoku Math. J. 19, 357-367.

Bosq, D. (1975). Inegalité de Bernstein pour les processus stationnaires et mélangeants, Applications, C.R. Acad. Sci. Paris, 281, A, 1095-1098.

Chow, Y. S. (1966). Some convergence theorems for independent random variables. Ann. Math. Statist. 37, 1482-1493.

Collomb, G. (1979). Conditions necessaires et sufficients de convergence uniform d'un estimateur de la régression, estimation des dérivées de la régression. C.R. Acad. Sci. Paris, 288, Ser. A, 161-164.

Collomb, G. (1984). Propriétès de convergence presque complete du prédicteur à noyau. Z. Wahr. verw.Gebiete 66, 441-460.

Collomb, G., and Härdle, W. (1986). Strong uniform convergence rates in robust nonparametric time series analysis and prediction: kernel regression estimation from dependent observations. Stoch. Proc. and its Application (to appear).

Deheuvels, P. (1974). Conditions necessaire et suffisantes de convergence ponctuelle presque sure et uniforme presque sure des estimateurs de la densite. C.R. Acad. Sci. Paris, Ser. A 278, 1217-1220.

Delecroix, M. (1980). Sur l'estimation des densités d'un processus stationaire a temps continue. Pulb. L'Inst. Statist. Univ. Paris 25, 17-40.

Devroye, L.P., Wagner, T.L. (1976). Nonparametric discrimination and density estimations, Technical Report 183, University of Texas, Austin.

Eberlein, E. (1986). On strong invariance principle under dependence assumptions, Ann. of Prob., 14, 260-271.

Kahane, J.P. (1986). Some random series of functions, Cambridge Studies in Advanced Mathematics 5.

Komlos, J., Major, P., and Tusnady, G. (1975). An approximation of partial sums of independent random variables and the sample distribution functions. I. Z. Wahr. verw. Gebiete 32, 111-131.

Härdle, W. and Marron, S. (1985). Optimal bandwidth selection in nonparametric regression function estimation, Ann. of Statistics·13, 1465-1482.

Ibragimov, I.A. (1962). Some limit theorem for stationary processes, Theory Prob. Appl. 7, 349-382.

Iosifescu, M., Theodorescu, R. (1969). Random Processes and Learning, Springer-Verlag.

Nadaraja, E. (1964). On regression estimators, Theory Probab. Appl. 9, 157-159.

Parzen, E. (1962). On estimation of a probability density function and mode, Ann. Math. Statist. 33, 1065-1076.

Peligrad, M. (1986). Properties of uniform consistency of the kernel estimator of a density and of a regression function under dependence assumptions. Preprint.

Philipp, W. (1977). A functional law of the iterated logarithm for empirical distributions functions of weakly dependent random variables, Ann. Probab. 5, 319-350.

Prakasa, Rao, B.L.S. (1983). Nonparametric functional estimation, Academic Press, New York.

Rosenblatt, M. (1956). Remarks on some nonparametric estimates of a density function, Ann. Math. Statist. 27, 832-837.

Rosenblatt, M. (1971). Markov processes, structure and asymptotic behavior, Springer-Verlag, Berlin.

Roussas, G.G. (1969). Nonparametric estimation in Markov processes, Ann. Inst. Statist. Math. 21, 73-87.

Rüschendorff, L. (1977). Consistency of estimators for multivariate density functions and for the mode. Sankhya Ser. A 39, 243-250.

Samur, J. (1984). Convergence of sums of mixing triangular arrays of random vectors with stationary rows, Ann. Probab. 12, 390-426.

Serfling, R.J. (1969). Probability inequalities and convergence properties for sums of multiplicative random variables, Rep. M151. Florida State Dept. of Statist., Tallahasse, Florida.

Silverman, B.W. (1978). Weak and strong uniform consistency of the kernel estimate of a density function and its derivatives, Ann. Statist. 6, 177-184 (Add. 8, 1175-1176 (1980)).

Stout, W. (1974). Almost sure convergence. Academic Press.

Watson, G.S. (1964). Smooth regression analysis, Sankhya Ser. A., 26, 359-372.

Yoshihara, K. (1978). Probability inequalities for sums of absolutely regular processes and their applications, Z. Wahr. verw. Gebiete, 43, 319-329.

# A LIMIT THEOREM FOR SOME MODIFIED CHI-SQUARE STATISTICS WHEN THE NUMBER OF CLASSES INCREASES

F. C. Drost
Dept. of Mathematics and Computer Science
Free University
De Boelelaan 1081
1081 HV Amsterdam
Holland

ABSTRACT. In the presence of a location-scale nuisance parameter we consider three chi-square type tests based on increasingly finer partitions as the sample size increases. The asymptotic distributions are derived both under the null-hypothesis and under local alternatives, obtained by taking contamination families of densities between the null-hypothesis and fixed alternative hypotheses. As a consequence of our main theorem it is shown that the Rao-Robson-Nikulin test asymptotically dominates the Watson-Roy test and the Dzhaparidze-Nikulin test. Conditions are given when it is optimal to let the number of classes increase to infinity.

*Key-words:* chi-square tests, location-scale parameter, goodness-of-fit, number of classes.

AMS 1980 *subject classification:* Primary 62E20, 62F20, secondary 62F05.

## 1. INTRODUCTION AND NULL-HYPOTHESIS THEORY

Let $Y_1,\ldots,Y_n$ be i.i.d. real-valued absolutely continuous random variables with distribution function $F^Y$ and consider the general testing problem $F^Y = F_0$ for a given distribution function $F_0$ (with density $f_0$). A well-known omnibus goodness-of-fit test is the classical Pearson chi-square test. In the presence of nuisance parameters (and k bounded) many modifications of this test have been proposed (cf. Roy (1956), Watson (1958, 1959), Nikulin (1973), Dzhaparidze and Nikulin (1974), Rao and Robson (1974), Moore and Spruill (1975)). In this paper we restrict attention to a location-scale parameter $\theta = (\mu,\sigma)'$.



*P. Bauer et al. (eds.), Mathematical Statistics and Probability Theory, Vol. B, 43–58.*

Suppose that the real line is partitioned into k cells

(1.1) $\quad I_{ki}(\theta) = (\mu + a_{ki-1}\sigma, \mu + a_{ki}\sigma] \qquad (i = 1,\ldots,k),$

where the constants $-\infty = a_{k0} < \ldots < a_{kk} = \infty$ are chosen such that the cells are equiprobable under $F_0^*(\cdot;\theta) = F_0((\cdot-\mu)/\sigma)$. Define

(1.2) $\quad p_{ki}(0) = \int_{I_{ki}(\theta)} d\, F_0^*(y;\theta) = \int_{(a_{ki-1},a_{ki}]} d\, F_0(y) = 1/k,$

(1.3) $\quad N_{ki}(\theta) = \#\{j;\ Y_j \in I_{ki}(\theta)\}$ and

(1.4) $\quad V_{ki}(\theta) = (N_{ki}(\theta) - np_{ki}(0))/(np_{ki}(0))^{\frac{1}{2}} \quad (i = 1,\ldots,k).$

When $\theta$ is known the Pearson test rejects $F^Y = F_0$ for large values of

(1.5) $\quad P_k = V_k'(\theta)V_k(\theta) = \sum_{i=1}^{k} V_{ki}^2(\theta).$

In the present case the nuisance parameter $\theta$ has to be estimated. A natural estimator is the maximum likelihood estimator $\hat{\theta}_n^{ML}$ based on the raw data. We assume that the Fisher-information-matrix $J_\theta = \sigma^2 J$ finitely exists and that it is invertible. Let $I_k$ be the (k×k)-indentity-matrix,
$q_k = (p_{k1}^{\frac{1}{2}}(0),\ldots,p_{kk}^{\frac{1}{2}}(0))$, $B_k$ the (k×2)- matrix with i-th rows

(1.6) $\quad B_{ki} = p_{ki}^{-\frac{1}{2}}(0)\nabla_\theta \int_{(a_{ki-1},a_{ki}]} d\, F^*(y;\theta)\Big|_{\theta=(0,1)'}$

and $\Sigma_k$ the (k×k)-matrix

(1.7) $\quad \Sigma_k = I_k - B_k J^{-1} B'_k.$

Note that $q'_k V_k(\theta) = 0$, $q_k' B_k = 0$ and $q_k' q_k = 1$. Assume that $\Sigma_k$ is invertible. Theorem 4.1 of Moore and Spruill (1975) shows that $V_k(\hat{\theta}_n^{ML}) \approx V_k(\theta) - B_k n^{\frac{1}{2}}\sigma^{-1}(\hat{\theta}_n^{ML} - \theta)$. The covariance matrix of the RHS is $\Sigma_k - q_k q_k'$ (cf. Remark 3.1). The Moore-Penrose generalized inverse of this covariance matrix is given by

$$(1.8)\qquad \Sigma_k^{-1} = I_k + B_k(J - B_k' B_k)^{-1} B_k' .$$

It follows from (1.7) that the matrix $J - B_k'B_k$ is positive definite. For the testing problem

$$(1.9)\qquad H_0 : F^Y \in \{F_0^*(\cdot;\theta);\ \mu \in \mathbf{R},\ \sigma > 0\}$$

the following chi-square type test statistics are studied (as $k \to \infty$):

I. The Watson-Roy statistic $WR_k = V_k'(\hat\theta_n^{ML})\ V_k(\hat\theta_n^{ML})$,

II. The Dzhaparidze-Nikulin statistic

$$DN_k = V_k'(\hat\theta_n^{ML})[I_k - B_k(B_k'B_k)^{-1}B_k']V_k(\hat\theta_n^{ML}),$$

III. The Rao-Robson-Nikulin statistic $RR_k = V_k'(\hat\theta_n^{ML})\ \Sigma_k^{-1}\ V_k(\hat\theta_n^{ML})$.

First we review some results for fixed k(cf. Moore and Spruill (1975) for an exhaustive treatment).Note (cf. also McCulloch (1985))

$$(1.10)\qquad DN_k \le WR_k \le RR_k .$$

The Watson-Roy statistic resembles the simple Pearson statistic but it has an awkward limiting null distribution

$$(1.11)\qquad WR_k \to_{d_0} \chi^2_{k-3} + \mu_1\chi^2_1 + \mu_2\chi^2_1 ,$$

where $0 \le \mu_1 \le \mu_2 \le 1$ are the eigenvalues of $\Sigma_k$ corresponding to eigenvectors in the column space of $B_k$ . The two other statistics are more difficult to calculate but have limiting $\chi^2$ distributions

$$(1.12)\qquad DN_k \to_{d_0} \chi^2_{k-3} \text{ and}$$

$$(1.13)\qquad RR_k \to_{d_0} \chi^2_{k-1} .$$

In this paper k slowly tends to infinity as $n \to \infty$. We show that the test statistics I - III have the same normal limiting null distributions with parameters k and 2k (cf. Bickel and Rosenblatt (1973) for a similar result about a modification of $WR_k$). For $k \to \infty$ these parameters are the leading terms of the expectations and variances of the limiting distributions for fixed k. The proof rewrites the statistics as the sum of the classical Pearson statistic and some remainder terms which are small in probability. The desired result then follows form Theorem 5.1 of Morris (1975). Because of (1.10) this immediately implies a remarkable result for Pitman efficiencies as $k \to \infty$

(1.14) $e_p(DN,WR) \le 1$ and

(1.15) $e_p(WR,RR) \le 1.$

For fixed k a similar relation is not true (cf. Moore(1977), Le Cam et al. (1983)). Example 5.1 illustrates that the order may even be reversed. In many cases, however, simulations are in accordance with (1.14) and (1.15) (cf. Stephens (1974)).

## 2. LOCAL ALTERNATIVES

To derive the limiting distributions under local alternatives let $F_1$ be a given alternative and consider the sequence of local alternative hypotheses

$$H_{1n}: F^Y \in \{F^*_{\eta_n}(\cdot;\theta) = F_{\eta_n}((\cdot-\mu)/\sigma) = (1-\eta_n)F_0((\cdot-\mu)/\sigma) + \eta_n F_1((\cdot-\mu)/\sigma);\ \mu \in \mathbf{R},\ \sigma > 0\}, \tag{2.1}$$

where $\eta_n \to 0$ as $n \to \infty$. A common choice for $\eta_n$ is

$$\eta_n = n^{-\frac{1}{2}}\gamma \tag{2.2}$$

for some fixed $\gamma > 0$, keeping the power bounded away from $\alpha$ and 1 for fixed k. Let $E_\eta$ denote the expectation with respect to $F^*_\eta(\cdot;\theta)$. Denote the expectation of the principle part of $\hat{\theta}_n^{ML}$ (cf. (3.8)) by

$$E = E_1\{h((Y_1 - \mu)/\sigma)\}, \tag{2.3}$$

let $p_k(\eta)$ be the k-vector of cell-probabilities under $F^*_\eta(\cdot;\theta)$

$$p_{ki}(\eta) = \int_{I_{ki}(\theta)} d\, F^*_\eta(y;\theta) = \int_{(a_{ki-1},a_{ki}]} d\, F_\eta(y) \qquad (i = 1,\ldots,k) \tag{2.4}$$

and let $d_k$ be the k-vector with components

$$d_{ki} = n^{\frac{1}{2}}\eta_n(p_{ki}(1) - p_{ki}(0))/\, p^{\frac{1}{2}}_{ki}(0) \qquad (i = 1,\ldots,k). \tag{2.5}$$

We denote the 'noncentrality parameters' corresponding to $WR_k$, $DN_k$ and $RR_k$ by

$$\delta_k^{WR} = (d_k - n^{\frac{1}{2}}\eta_n B_k E)'\,(d_k - n^{\frac{1}{2}}\eta_n B_k E), \tag{2.6}$$

$$\delta_k^{DN} = (d_k - n^{\frac{1}{2}}\eta_n B_k E)'[I_k - B_k(B_k'\, B_k)^{-1}B_k']\,(d_k - n^{\frac{1}{2}}\eta_n B_k E) \text{ and} \tag{2.7}$$

$$\delta_k^{RR} = (d_k - n^{\frac{1}{2}}\eta_n B_k E)'\Sigma_k^{-1}\,(d_k - n^{\frac{1}{2}}\eta_n B_k E). \tag{2.8}$$

For fixed k the limiting alternative distributions are derived from Theorem 4.2 of Moore and Spruill (1975). Let $w_{k1}$ and $w_{k2}$ be the eigenvectors of $\Sigma_k$ with eigenvalues $\mu_1$ and $\mu_2$ and let
$v_j = w_{kj}'(d_k - n^{\frac{1}{2}}\eta_n B_k E) \quad (j = 1,2)$.
Then

$$WR_k \to_{d_{1n}} \chi'^2_{k-3}(\delta_k^{WR} - v_1^2 - v_2^2) + \mu_1\chi_1'^2\,(v_1^2/\mu_1) + \mu_2\chi_1'^2\,(v_2^2/\mu_2), \tag{2.9}$$

$$DN_k \to_{d_{1n}} \chi'^2_{k-3}(\delta_k^{DN}) \text{ and} \tag{2.10}$$

$$RR_k \to_{d_{1n}} \chi'^2_{k-1}(\delta_k^{RR}). \tag{2.11}$$

Let $k \to \infty$. As under $H_0$, the limiting alternative distributions are the first order approximations of the limiting distributions for fixed k (cf. the proof of Theorem 4.1). They are normal with parameters $k + \delta_k$ and 2k (Theorem 4.1). As a consequence of Theorem 4.1 we can calculate the Pitman efficiencies (extending (1.14) and (1.15))

$$(2.12) \qquad e_p(\mathcal{DN},\mathcal{WR}) = \lim_{k\to\infty} \delta_k^{DN} / \delta_k^{WR} \qquad (\le 1) \quad \text{and}$$

$$(2.13) \qquad e_p(\mathcal{WR},\mathcal{RR}) = \lim_{k\to\infty} \delta_k^{WR} / \delta_k^{RR} \qquad (\le 1).$$

In many examples the inequality is strict (cf. Example 4.3 of Drost (1986)). Another consequence is similar to the one obtained by Kallenberg et al. (1985), who derived a criterion for the Pearson test (when no nuisance parameters are present) whether it is better to keep k bounded or to let $k \to \infty$. We obtain for each of the tests I - III (cf. Corollary 4.3 and Remark 4.3)

$$(2.14) \qquad \delta_k/k^{\frac{1}{2}} \to \begin{cases} 0 \\ \infty \end{cases} \Rightarrow \begin{matrix} \text{asymptotic local power} \\ \text{is highest for} \end{matrix} \begin{cases} \text{small } k \\ k \to \infty \end{cases}.$$

## 3. ASSUMPTIONS

For simplicity we impose similar conditions under the local alternatives $H_{1n}$ as under $H_0$ although they can often be relaxed (cf. Drost (1986)). Throughout the rest of the paper all these conditions are implicitly assumed with the exception of condition C.

### 3.1. Assumptions on the distributions

Assume that the densities $f_0$ and $f_1$ of $F_0$ and $F_1$ are Lipschitz continuous with Lipschitz constant L

$$(3.1) \qquad \forall x,y \in \mathbb{R} \qquad |f_j(x) - f_j(y)| \le L|x - y| \qquad (j = 0,1)$$

and assume

(3.2) $$\lim_{|x|\to\infty} xf_0(x) = 0.$$

Furthermore we assume that $F_1$ is absolutely continuous with respect to $F_0$, implying

(3.3) $$\lim_{k\to\infty} \max_{1\le i\le k} p_{ki}(1) = 0 \quad \text{and}$$

(3.4) $$d_k' d_k = \mathcal{O}(k) \text{ as } k \to \infty.$$

3.2. Assumptions on the rate of k

Let $k = k(n)$ $(n \in \mathbb{N})$ be a particular sequence for the number of cells and assume, as $n \to \infty$,

(3.5) $$k \to \infty,$$

(3.6) $$k^2 \log^{\frac{3}{2}} k = o(n^{\frac{1}{2}}) \quad \text{and}$$

(3.7) $$\max(a_{k1}^4, a_{kk-1}^4) = \mathcal{O}(n^{\frac{1}{2}} \log^{\frac{3}{2}} k).$$

For the Rao-Robson-Nikulin test we need an additional condition. Let $\lambda_k$ denote the maximum eigenvalue of $\Sigma_k^{-1}$ and assume

C $$\lambda_k k^{\frac{3}{2}} \log^{\frac{3}{2}} k = o(n^{\frac{1}{2}}).$$

In special cases one can show that $\lambda_k$ is approximately of order k (cf. Example 5.1).

3.3. Assumptions on the estimator $\hat{\theta}_n^{ML}$

Under regularity conditions Bickel (1982) showed that $\hat{\theta}_n^{ML}$ admits the representation

(3.8) $$n^{\frac{1}{2}}\sigma^{-1}(\hat{\theta}_n^{ML} - \theta) = n^{-\frac{1}{2}} \sum_{j=1}^{n} h((Y_j - \mu)/\sigma) +$$

$$+ Q_n((Y_1-\mu)/\sigma,\ldots,(Y_n-\mu)/\sigma),$$

where the remainder $Q_n: \mathbb{R}^n \to \mathbb{R}^2$ is small in probability (often of order $\mathcal{O}_p(n^{-\frac{1}{2}})$) and where $h: \mathbb{R} \to \mathbb{R}^2$ is the influence function

$$\text{(3.9)}\qquad \begin{aligned} h(y) &= J^{-1}\nabla_\theta \ln f_0^*(y;\theta)\Big|_{\theta=(0,1)'} \\ &= -J^{-1}(f_0'(y)/f_0(y), 1 + yf_0'(y)/f_0(y))'. \end{aligned}$$

We omit these regularity conditions but assume that the representation holds with

$$\text{(3.10)}\qquad Q_n = \mathcal{O}_p(n^{-\frac{1}{4}}) \qquad \text{under } H_0 \text{ and } H_{1n}.$$

Note that $E_0\{h((Y_1 - \mu)/\sigma)\} = 0$, $E_0[h((Y_1 - \mu)/\sigma)h'((Y_1 - \mu)/\sigma)] = J^{-1}$ and

$$\text{(3.11)}\qquad B_{ki} = \int_{(a_{ki-1}, a_{ki}]} h'(y)d\,F_0(y)\cdot J \qquad (i = 1,\ldots,k).$$

We also assume that the first two moments of h finitely exist under $H_1$

$$\text{(3.12)}\qquad E_1\{h'((Y_1 - \mu)/\sigma)h((Y_1 - \mu)/\sigma)\} < \infty.$$

REMARK 3.1. Using (3.8) and Theorem 4.1 of Moore and Spruill (1975) the covariance matrix of the principle part of $V_k(\hat{\theta}_n^{ML})$ is obtained from

$$\text{(3.13)}\qquad \operatorname{cov}_0\{V_k(\theta) - B_k n^{-\frac{1}{2}} \sum_{j=1}^{n} h((Y_j - \mu)/\sigma)\} = \Sigma_k - q_k q_k'. \qquad \square$$

## 4. MAIN RESULTS

In this section we assume without further reference the conditions of Sections 1 - 3. Our main theorem concerns the limiting distributions of $WR_k$, $DN_k$ and $RR_k$ both under $H_0$ and $H_{1n}$.

THEOREM 4.1. *Consider the chi-square type tests* I - III *of* $H_0$ *versus the sequence of local hypotheses* $H_{1n}$ *determined by* $F_1$. *Then*

(4.1a) $$(WR_k - k)/(2k)^{\frac{1}{2}} \to_{d_0} N(0,1) \quad \text{and}$$

(4.1b) $$(WR_k - (k + \delta_k^{WR}))/(2k)^{\frac{1}{2}} \to_{d_{1n}} N(0,1).$$

(4.2a) $$(DN_k - k)/(2k)^{\frac{1}{2}} \to_{d_0} N(0,1) \quad \text{and}$$

(4.2b) $$(DN_k - (k + \delta_k^{DN}))/(2k)^{\frac{1}{2}} \to_{d_{1n}} N(0,1).$$

*Assume the additional condition C, then*

(4.3a) $$(RR_k - k)/(2k)^{\frac{1}{2}} \to_{d_0} N(0,1),$$

(4.3b) $$(RR_k - (k + \delta_k^{RR}))/(2k)^{\frac{1}{2}} \to_{d_{1n}} N(0,1) \text{ if}$$

$$\limsup_{k\to\infty} (\delta_k^{RR} - \delta_k^{WR})/k^{\frac{1}{2}} < \infty \text{ and}$$

(4.3c) $$(RR_k - (k + \delta_k^{WR}))/(2k)^{\frac{1}{2}} \to_{p_{1n}} \infty \text{ if } \lim_{k\to\infty} (\delta_k^{RR} - \delta_k^{WR})/k^{\frac{1}{2}} = \infty.$$

PROOF. For the formal proof we refer to Drost (1986). Here the general lines of the proof are indicated followed by an intuitive argument.

First rewrite the stochastic k-vector $V_k(\hat{\theta}_n^{ML})$ as in Moore and Spruill (1975)

$$V_k(\hat{\theta}_n) = V_k(\theta) - B_k\, n^{-\frac{1}{2}} \sum_{j=1}^{n} h((Y_j - \mu)/\sigma) + \text{remainder}.$$

The analysis of the first part heavily relies on Theorem 5.1 of Morris (1975). The stochastic remainder causes more difficulties because its dimension grows with k. To bound the remainder we need conditions like (3.6), (3.7), (3.10) and C (generally it is not true that the mean of k tight random variables is tight). More general but less transparent conditions are given in Drost (1986).

A sound impression of the limiting distributions (4.1) - (4.3) for $k\to\infty$ is obtained from the limiting distributions (2.9) - (2.11) for fixed k. Note that the latter distributions are (approximately) sums of independent identical noncentral $\chi_1^2$ variables with expectations

$$E\{\chi'^2_{k-3}(\delta_k^{WR} - v_1^2 - v_2^2) + \mu_1\chi'^2_1(v_1^2/\mu_1) + \mu_2\chi'^2_1(v_2^2/\mu_2)\} = k - 3 + \mu_1 + \mu_2 + \delta_k^{WR}$$

$$E\{\chi'_{k-3}(\delta_k^{DN})\} = k - 3 + \delta_k^{DN}$$

$$E\{\chi'_{k-1}(\delta_k^{RR})\} = k - 1 + \delta_k^{RR}$$

and variances

$$\text{var}\{\chi'^2_{k-3}(\delta_k^{WR} - v_1^2 - v_2^2) + \mu_1\chi'^2_1(v_1^2/\mu_1) + \mu_2\chi'^2_1(v_2^2/\mu_2)\}$$

$$= 2(k - 3 + \mu_1^2 + \mu_2^2) + 4(\delta_k^{WR} + (\mu_1 - 1)v_1^2 + (\mu_2 - 1)v_2^2)$$

$$\text{var}\{\chi'^2_{k-3}(\delta_k^{DN})\} = 2(k - 3) + 4\delta_k^{DN}$$

$$\text{var}\{\chi'^2_{k-1}(\delta_k^{RR})\} = 2(k - 1) + 4\delta_k^{RR}.$$

Using (3.4) the expectations for fixed k are seen to be equal to $k + \delta_k + o(k^{\frac{1}{2}})$ and the variances are equal to $2k + o(k)$. Suppose that the convergence in distribution for fixed k is uniformly in k, then the result follows. □

REMARK 4.1. Obviously Theorem 4.1 continues to hold if $\theta$ is either a location or a scale parameter. □

REMARK 4.2. Bickel and Rosenblatt (1973) obtained a similar result for a modification of $WR_k$ under slightly different conditions (cf. also Chapter 6 of Csörgö and Révész (1981)). Their result is not directly applicable to $DN_k$ and $RR_k$. The cross-terms

$V_{ki}(\hat{\theta}_n^{ML})\ V_{kj}(\hat{\theta}_n^{ML})$ (i,j = 1,...,k) appearing in these statistics seriously complicate the proof. □

In the remainder of this section we state a corollary concerning the relative efficiency of the test statistics I - III and another about the number of classes. First we introduce some notation.
For each $k \geq 2$ denote the power under $H_{1n}$ of a test statistic $S_k$ which is based on k random cells and n observations by

(4.4) $$\beta_\alpha(S_k,n,\eta_n) = P_{1n}(S_k(Y_1,\ldots,Y_n) > c_k),$$

where the critical values $c_k$ are given by

(4.5) $$c_k = \inf\{c;P_0(S_k(Y_1,\ldots,Y_n) \geq c) \leq \alpha\}.$$

Let $S_k^*$ $(k \geq 2)$ be some other sequence of test statistics and define the sequence $n_1(n)$

(4.6) $$n_1(n) = \min\{n_1;\beta_\alpha(S_k,n,\eta_n) - \beta_\alpha(S^*_{k(n_1)},n_1,n_n) \leq 0\}.$$

The Pitman efficiency of $S_k$ with respect to $S_k^*$ is defined by

(4.7) $$e_p(S,S^*) = \lim_{n\to\infty} n_1(n)/n$$

provided that this limit exists. In Corollary 4.2 the Pitman efficiencies of the test statistics under consideration are investigated. The Rao-Robson-Nikulin statistic turns out to be most efficient. This is supported by simulation studies of Stephens (1974) and a result on the approximate Bahadur slope by Spruill (1976) (for k fixed).

COROLLARY 4.2.
(i) *Assume that* $\delta_k^{DN}/k^{\frac{1}{2}}$ *has a finite limit and that* $\delta_k^{WR}/k^{\frac{1}{2}}$ *has a positive limit, then*

(4.8) $$e_p(DN,WR) = \lim_{k\to\infty} \delta_k^{DN}/\delta_k^{WR} \quad (\leq 1).$$

ii) *Assume* C *and assume that* $\delta_k^{WR}/k^{\frac{1}{2}}$ *has a finite limit and that* $\delta_k^{RR}/k^{\frac{1}{2}}$ *has a positive limit, then*

$$(4.9) \qquad e_p(WR,RR) = \lim_{k\to\infty} \delta_k^{WR}/\delta_k^{RR} \qquad (\leq 1).$$

PROOF. Because of (4.1) - (4.3) the asymptotic local power is completely determined by the behaviour of the ratio $\delta_k/k^{\frac{1}{2}}$. This ratio essentially depends on n through $n\eta_n^2$. The conclusion of the corollary then folows from (4.7) and

$$n\eta_n^2\ (\delta_{k(n)}^{DN}/\gamma^2)/k^{\frac{1}{2}}(n) = n_1(n)\eta_n^2\ (\delta_{k(n_1)}^{WR}/\gamma^2)/k^{\frac{1}{2}}(n_1) \qquad \text{and}$$

$$n\eta_n^2\ (\delta_{k(n)}^{WR}/\gamma^2)/k^{\frac{1}{2}}(n) = n_1(n)\eta_n^2\ (\delta_{k(n_1)}^{RR}/\gamma^2)/k^{\frac{1}{2}}(n_1). \qquad \square$$

Corollary 4.3 is similar to Proposition 4.1 of Kallenberg et al. (1985). They proved for the Pearson statistic and $k \to \infty$ that the ratio of the noncentrality parameter and the square root of k determines the asymptotic local power. Along the same lines one shows that the same result is true for $WR_k$, $RR_k$ and $DN_k$.

COROLLARY 4.3.

$$\lim_{n\to\infty} \beta_\alpha(WR_k,n,\eta_n) = \begin{cases} \alpha \\ \\ 1 \end{cases} \text{iff} \lim_{k\to\infty} \delta_k^{WR}/k^{\frac{1}{2}} = \begin{cases} 0 \\ \cdot \\ \infty \end{cases}$$

$$\lim_{n\to\infty} \beta_\alpha(DN_k,n,\eta_n) = \begin{cases} \alpha \\ \\ 1 \end{cases} \text{iff} \lim_{k\to\infty} \delta_k^{DN}/k^{\frac{1}{2}} = \begin{cases} 0 \\ \cdot \\ \infty \end{cases}$$

*Assume the additional condition* C, *then*

$$\lim_{n\to\infty} \beta_\alpha(RR_k,n,\eta_n) = \begin{cases} \alpha \\ \\ 1 \end{cases} \text{iff} \lim_{k\to\infty} \delta_k^{RR}/k^{\frac{1}{2}} = \begin{cases} 0 \\ \cdot \\ \infty \end{cases} \qquad \square$$

REMARK 4.3. For bounded k the choice $\eta_n = n^{-\frac{1}{2}}\gamma$ results in an asymptotic local power between $\alpha$ and 1. Thus Corollary 4.3 implies (2.14). □

## 5. AN EXAMPLE

The example shows that, for fixed k, one cannot order the statistics under consideration. The order which holds for $k \to \infty$ may even be reversed.

EXAMPLE 5.1. Consider the testing problem of a normal null-hypothesis with unknown location versus a regular symmetric alternative. Then, for fixed k,

$$(5.1) \qquad e_p(DN,WR) > 1 \qquad \text{and}$$

$$(5.2) \qquad e_p(WR,RR) > 1,$$

while, for $k \to \infty$ and $k^{\frac{5}{2}} \log^{\frac{5}{2}} k = o(n^{\frac{1}{2}})$,

$$(5.3) \qquad e_p(DN,WR) = e_p(WR,RR) = 1.$$

To prove (5.1) - (5.3) we first present a lemma of independent interest, which extends a result in Section 3.4 of Lehmann (1959).

LEMMA 5.1. *Let* X *and* Y *be independent random variables. Suppose that under* $H_0$ $(H_1)$ X *has distribution* $F_0^X$ $(F_1^X)$ *with density* $f_0^X$ $(f_1^X)$ *and assume that the ratio* $f_1^X(x)/f_0^X(x)$ *is strictly increasing in* x. *Suppose that the distribution* $F^Y$ *of* Y *is independent of* $H_0$ *or* $H_1$. *Consider test statistics* $Z_\lambda = X + \lambda Y$ *and reject* $H_0$ *for large values. If* $\lambda < \mu \le 0$ *or if* $0 \le \mu < \lambda$ *then* $Z_\lambda$ *is strictly less informative than* $Z_\mu$ (cf. Lehmann (1959) p.75).

PROOF. Let $c_{\alpha,\lambda} = \inf\{c; P_0(Z_\lambda \ge c) \le \alpha\}$ and $\beta_\lambda = P_1(Z_\lambda > c_{\alpha,\lambda})$. Let (s,t) be the solution of

$$\begin{cases} x + \lambda y = c_{\alpha,\lambda} \\ \\ x + \mu y = c_{\alpha,\mu} \end{cases}$$

then

$$\beta_\lambda = P_1(Z_\lambda > c_{\alpha,\lambda})$$

$$= P_1(Z_\mu > c_{\alpha,\mu}) + \int_{\substack{x+\lambda y>c_{\alpha,\lambda}\\ x+\mu y\le c_{\alpha,\mu}}} dF_1^X(x)\, dF^Y(y) - \int_{\substack{x+\lambda y\le c_{\alpha,\lambda}\\ x+\mu y>c_{\alpha,\mu}}} dF_1^X(x)\, dF^Y(y)$$

$$< \beta_\mu + f_1^X(s)/f_0^X(s)\left\{\int_{\substack{x+\lambda y>c_{\alpha,\lambda}\\ x+\mu y\le c_{\alpha,\mu}}} dF_0^X(x)\, dF^Y(y) - \int_{\substack{x+\lambda y\le c_{\alpha,\lambda}\\ x+\mu y>c_{\alpha,\mu}}} dF_0^X(x)\, dF^Y(y)\right\}$$

$$= \beta_\mu. \qquad \square$$

Because of the symmetry in the example $d_k' B_k = 0$ and $E = 0$. The limiting distributions for fixed k follow from Theorem 4.2 of Moore and Spruill (1975)

$$DN_k \to_{d_0} \chi^2_{k-2},$$

$$WR_k \to_{d_0} \chi^2_{k-2} + (1 - B_k' B_k)\chi_1^2,$$

$$RR_k \to_{d_0} \chi^2_{k-2} + \chi_1^2,$$

$$DN_k \to_{d_{1n}} \chi'^2_{k-2}(d_k' d_k),$$

$$WR_k \to_{d_{1n}} \chi'^2_{k-2}(d_k' d_k) + (1 - B_k' B_k)\chi_1^2 \quad \text{and}$$

$$RR_k \to_{d_{1n}} \chi'^2_{k-2}(d_k' d_k) + \chi_1^2.$$

Lemma 5.1 and $1 - B_k' B_k > 0$ imply (5.1) and (5.2). Note that the conditions of Sections 1 - 3 are satisfied (use $\lambda_k = 1/(1 - B_k' B_k) = \mathcal{O}(k \log k)$ ). Corollary 4.2 then yields (5.3). In this example normality is not essential, symmetry suffices.

REFERENCES

[1] BICKEL, P.J. (1982), 'On adaptive estimation', *Ann. Statist.* **10**, 647-671.

[2] BICKEL, P.J. and ROSENBLATT, M. (1973), 'On some global measures of the deviation of density function estimates', *Ann. Statist.* **1**, 1071-1095.

[3] CSÖRGŐ, M. and RÉVÉSZ, P. (1981), *Strong Approximations in Probability and Statistics*, Academic Press.

[4] DROST, F.C. (1986), 'Generalized chi-square goodness-of-fit tests for location-scale models when the number of classes tends to infinity', Technical report 309, Dept. of Mathematics and Computer Science, Free University, Amsterdam.

[5] DZHAPARIDZE, K.O. and NIKULIN, M.S. (1974), 'On a modification of the standard statistics of Pearson', *Theory of probability and its Applications* **19**, 851-853.

[6] KALLENBERG, W.C.M., OOSTERHOFF, J. and SCHRIEVER, B.F. (1985), 'The number of classes in chi-squared goodness-of-fit tests', *J. Amer. Statist. Assoc.* **80**, 959-968.

[7] LECAM, L., MAHAN, C.M. and SINGH, A. (1983), 'An extension of a theorem of H. Chernoff and E.L. Lehmann', in: RIZVI, M.H. RUSTAGI, J.S. and SIEGMUND, D. (ed's), *Recent Advances in Statistics*, Academic Press, New York, 303-337.

[8] LEHMANN, E.L. (1959), *Testing Statistical Hypotheses*, Wiley, New York.

[9] McCULLOCH, C.E. (1985), 'Relationships among some chi-square goodness of fit statistics', *Commun. Statist. - Theor. Math.* **14** (3), 593-603.

[10] MOORE, D.S. (1977), 'Generalized inverses, Wald's method, and the construction of chi-squared tests of fit', *J. Amer. Statist. Assoc.* **72**, 131-137.

[11] MOORE, D.S. and SPRUILL, M.C. (1975), 'Unified large-sample theory of general chi-squared statistics for tests of fit', *Ann. Statist.* **3**, 599-616.

[12] MORRIS, C. (1975), 'Central limit theorems for multinomial sums', *Ann. Statist.* **3**, 165-188.

[13] NIKULIN, M.S. (1973), 'Chi-square test for continuous distributions with shift and scale parameters', *Theory of Probability and its Applications* **3**, 559-668.

[14] RAO, K.C. and ROBSON, D.S. (1974), 'A chi-square statistic for goodness of fit tests within the exponential family', *Comm. Statist.* **3**, 1139-1153.

[15] ROY, A.R. (1956), 'On $\chi^2$ statistics with variable intervals', Technical report No. 1, Stanford Univ. Dept. of Statistics.

[16] SPRUILL, M.C. (1976), 'A comparison of chi-square goodness-of-fit tests based on approximate Bahadur slope', *Ann. Statist.* **4**, 409-412.

[17] STEPHENS, M.A. (1974), 'EDF statistics for goodness of fit and some comparisons', *J. Amer. Statist. Assoc.* **69**, 703-737.

[18] WATSON, G.S. (1957), 'The $\chi^2$ goodness-of-fit test for normal distributions', *Biometrika* **44**, 336-348.

[19] WATSON, G.S. (1958), 'On chi-square goodness-of-fit tests for continuous distributions', *J.R. Statist. Soc. Ser. B* **20**, 44-61.

PROBLEMS AND RESULTS ON RANDOM WALKS

P.Erdös
Mathematical Institute
Reáltanoda u 13-15
1053 Budapest
Hungary

P.Rêvész
Mathematical Institute Budapest
and Technical University
Wiedner Hauptstraße 8-10/107
A-1040 Vienna, Austria

ABSTRACT: In this papes we present a number of unsolved problems of the simple, symmetric random walk together with the relevant known results.

## 1. INTRODUCTION

We consider the simple, symmetric random walk on the r-dimensional integer lattice. It is perhaps surprising how many unsolved problems remain in this old subject. In this paper we present a number of unsolved problems together with the relevant known results. We do not give any proofs but we give as many references as possible. Together with the presented unsolved problems we try to indicate whether we believe that they can be solved by the methods standing at our disposal or we feel that some new ideas of methods are necessary to settle them.

## 2. RANDOM WALK ON THE LINE

Let $X_1, X_2, \ldots$ be a sequence of i.i.d.r.v.'s with

$$P(X_1=+1) = P(X_1=-1) = 1/2$$

and

$$S_0 = 0, \quad S_n = \sum_{i=1}^{n} X_i \qquad (n=1,2,\ldots).$$

$S_n$ is considered as the location of the particle (involved in the random walk) after n steps.

### 2.1. The favourite values of a random walk

Let

$$\xi(x,n) = \#\{k: \ 0 \leq k \leq n, \ S_k = x\}$$

be the local time of the random walk, i.e. $\xi(x,n)$ is the number of visits in x up to n. A point $x_n$ is called a favourite value at the moment n if the particle visits $x_n$ most often during the first n steps i.e.



P. Bauer et al. (eds.), Mathematical Statistics and Probability Theory, Vol. B, 59–65.

$$\xi(x_n,n) = \max_x \xi(x,n).$$

The investigation of the properties of the favourite values started simultaneously by Bass and Griffin (1985) and ourselves (1984). One can easily observe that for infinitely many n there are two favourite values and also for infinitely many n there is only one favourite value with probability one. More formally speaking let $F_n$ be the set of favourite values i.e.

$$F_n = \{x: \xi(x,n) = \max \xi(x,n)\}$$

and let $|F_n|$ be the cardinality of $F_n$. Then

$$P\{|F_n| = 2 \text{ i.o.}\} = P\{|F_n| = 1 \text{ i.o.}\} = 1.$$

1.) We do not know wether 3 or more favourite values can occur infinitely often i.e. we ask:

$$P\{|F_n| = r \text{ i.o.}\} = ? \qquad (r=3,4,5,\ldots).$$

We thought that 0 is a favourite value i.o. that is $P\{0 \in F_n \text{ i.o.}\} = 1$. To our great surprise Bass and Griffin showed that it is not so and they proved that the favourite values are going to infinity faster than $n^{1/2}(\log n)^{-11}$. In fact they have

$$P\{\lim_{n\to\infty} \frac{(\log n)^{\alpha}}{n^{1/2}} \inf\{|x|, x \in F_n\} = \infty\} = 1$$

if $\alpha>11$. We showed that the favourite value i.o. larger than $(1-\varepsilon)(2n \log\log n)^{1/2}$ i.e.

$$P\{\limsup_{n\to\infty} (1+\varepsilon)(2n \log\log n)^{-1/2} \inf\{|x|, x\in F_n\}=1 \text{ i.o.}\}=1.$$

2.) We do not know whether the $\varepsilon$ can be replaced by 0 in the above statement.

Let $\alpha(n)$ be the number of different favourite values up to n, i.e.

$\alpha(n) = |\sum_{k=1}^{n} F_k|$. We guess that $\alpha(n)$ is very small i.e. $\alpha(n)< (\log n)^c$ for some $c > 0$ but we cannot prove it. Hence we ask

3.) How can one describe the limit behaviour of $\alpha(n)$?

4.) We also ask how long can a point stay as a favourite value i.e. let $1 \leqq i=i(n) < j=j(n) \leqq n$ be two integers for which

$$\left|\prod_{k=i}^{j} F_k\right| \geqq 1$$

and $j-i=\beta(n)$ is as big as possible. The question is to describe the limit behaviour of $\beta(n)$.

5.) Further if x was a favourite value once, can it happen that the favourite value moves away from x but later it returns to x again, i.e. do sequences $a_n<b_n<c_n$ of positive random integers exist such that

$$F_{a_n} F_{b_n} = \emptyset \quad \text{and} \quad F_{a_n} F_{c_n} \neq \emptyset \qquad (n=1,2,\ldots)?$$

6.) To investigate the jumps of the favourite values looks also interesting. Let $n=n(\omega)$ be a positive integer for which $F_n F_{n+1} = \emptyset$. Then the jump $j_n$ is defined as

$$j_n = \rho(F_n,F_{n+1}) = \min\{|x-y|;\ x \in F_n,\ y \in F_{n+1}\}.$$

The theorem of Bass and Griffin implies that $j_n \geqq n^{1/2}(\log n)^{-11}$ i.o.a.s.

It looks very likely that $\lim_{n\to\infty} j_n = \infty$ a.s. We do not see how one can describe the limit behaviour of $j_n$.

## 2.2 Long head-runs

We (1976) studied the length of the longest head-run $Z_n$, i.e. $Z_n$ is the largest integer for which

$$I(n,Z_n)=Z_n$$

where

$$I(n,k) = \max_{0\leq j\leq n-k} (S_{j+k}-S_j) \qquad (0 \leq k \leq n).$$

Our 1976 paper contained a complete enough characterization of $Z_n$ but the result was extended by Guibas-Odlyzko (1980), Samarova (1981), Révész (1982), Ortega-Wschebor (1984), Deheuvels (1985), Deheuvels, Devroye-Lynch (1986), Deheuvels-Steinebach (1986), Erdös-Révész (1986) among others. Now we propose some further problems.

Let $Z_n^*$ be the length of the longest tail run, i.e. $Z_n^*$ is the largest integer for which

$$I^*(n,Z_n^*) = -Z_n^*$$

where

$$I^*(n,k) = \min_{0\leq j\leq n-k} (S_{j+k}-S_j).$$

7.) How can we characterize the limit properties of $|Z_n-Z_n^*|$?

A trivial argument shows $P\{Z_n=Z_n^* \text{ i.o.}\}=1$ but it is not clear at all how big $|Z_n-Z_n^*|$ can be.

Let $Z_n^{(1)} = Z_n$ and let $Z_n^{(2)}, Z_n^{(3)}, \ldots$ be the length of the second, third,... longest head-run up to n.

8.) We ask about the properties of $Z_n^{(1)}-Z_n^{(2)}$. It is clear again that $P(Z_n^{(1)}=Z_n^{(2)} \text{ i.o.}) = 1$. The lim sup properties of $Z_n^{(1)}-Z_n^{(2)}$ look harder.

9.) Let $k_n$ be the largest integer for which

$$P\{Z_n^{(1)}=Z_n^{(2)}=\ldots=Z_n^{(k_n)} \text{ i.o.}\} = 1.$$

Characterize the limit properties of $k_n$.

## 2.3. On logarithmic and other densities

In many problems on random walk one gets density results only if we replace ordinary density by some more general concept of density. As an example we consider the sequence of time points when the particle returns to the origin. Let

$$Y_k = \begin{cases} 1 & \text{if} \quad S_k=0, \\ 0 & \text{if} \quad S_k\neq 0. \end{cases}$$

Then the sequence $\xi(0,n) = \sum_{k=1}^{n} Y_k$ does not obey law of large numbers but Chung and Erdös (1951) proved

$$\lim_{n\to\infty} (\log n)^{-1} \sum_{k=1}^{n} k^{-1/2} \; Y_k = \pi^{1/2} \quad \text{a.s.}$$

A similar result is due to P.Lévy who investigated the logarithmic density of those n's for which $S_n>0$. In fact let

$$V_n = \begin{cases} 1 & \text{if} \quad S_n > 0, \\ 0 & \text{if} \quad S_n \leqq 0. \end{cases}$$

Then P.Lévy proved that

$$(\log n)^{-1} \sum_{k=1}^{n} k^{-1} V_k = 1/2 \quad \text{a.s.}$$

In connection with result we ask:

10.) Does the sequence

$$(\log n)^{-1/2} \left( \sum_{k=1}^{n} k^{-1} V_k - \frac{1}{2} \log n \right)$$

satisfy the central limit theorem. This question does not seem to be very hard.

Our next problem is connected to the problem of long head-runs. Using the same notation as above let

$$U_n = \begin{cases} 0 & \text{if} \quad Z_n < Z_n^*, \\ 1 & \text{if} \quad Z_n > Z_n^*. \end{cases}$$

i.e. $U_n = 1$ if the longest head run up to n is longer than the longest tail run.

11.) Does the logarithmic density

$$\lim_{n\to\infty} (\log n)^{-1} \sum_{k=1}^{n} k^{-1} U_k$$

exist with probability one. This problem seems to be not very hard.

2.4. Rarely visited points

It is easy to see that for infinitely many n almost all paths assume every value at least twice which they assume at all, i.e. let

$$f_r(n) = \# \{k: \xi(k,n) = r\}$$

be the number of points visited exactly r-times up to n. Then

$$P\{f_1(n) = 0 \text{ i.o.}\} = 1.$$

12.) We do not know if for inifinitely many n almost all paths assume every value at least (r+1)-times (r=2,3,...) which they assume at all, i.e. let

$$\sum_{j=1}^{r} f_j(n) = g_r(n)$$

and we ask

$$P\{g_r(n) = 0 \text{ i.o.}\} = ?$$

We would guess that this probability is 0 if r > 2 but perhaps it is 1 if r=2.

13.) For every r (r may depend on n) investigate

$$\liminf_{n\to\infty} f_r(n) \text{ and } \limsup_{n\to\infty} f_r(n).$$

As already stated $\liminf_{n\to\infty} f_1(n) = 0$. P.Major (1986) proved

$$\limsup_{n\to\infty} \frac{f_1(n)}{\log^2 n} = C \qquad \text{a.s.}$$

where $0 < C < \infty$ but its exact value is unknown.
Let $e_i$ be the i-th unit-vector on $R^d$ i.e. $e_i=(0,0,\ldots 0,i,0,\ldots 0)$ and let $X_1, X_2,\ldots$ be a sequence of i.i.d.r.v.'s with

$$P\{X_1=e_i\} = P\{X_1=-e_i\} = \frac{1}{2d} \qquad (i=1,2,\ldots,d).$$

Further let

$$S_0=0, \quad S_n=X_1+X_2+\ldots+X_n \qquad (n=1,2,\ldots).$$

## 3. RANDOM WALK IN THE SPACE

Most of the problems formulated in Section 2 can be reformulated in d-dimension and a number of new problems can be found. At first we give a few remarks to the already stated problems. Later we present some new problems.

### 3.1. Multivariate versions of the one dimensional problems.

In connection with the favourite values it is natural to ask

14.) Does the favourite value of the random walk converge to inifinity in case d=2?

The answer is clearly positive when $d \geq 3$ and very likely it is also so in case d=2 but the proof is not clear.

In connection with the rarely visited points a result of Dvoretzky and Erdös (1950) implies that in case $d \geq 2$ a.s. there will be many points visited exactly once if n is big enough. In fact we have $\lim_{n\to\infty} f_r(n) = \infty$ a.s. if $d \geq 2$. Dvoretzky and Erdös as well as Erdös and Taylor (1960 and 1960) have some results to describe the limit properties of $f_r(n)$ but a complete description is missing.

### 3.2. Special problems in case $d \geq 2$.

Let us consider the largest square around the origin completely covered by the path during the first n steps. Clearly we say that a square $[-A,A] \times [-A,A]$ is completely covered during the first n steps if for any $x \in [-A,A] \times [-A,A]$ there exists a $1 \leq k = k_n(x) \leq n$ such that $S_k=x$. Let $A_n$ be the largest integer for which the square $[-A_n,A_n] \times [-A_n,A_n]$ is completely covered. Clearly $\lim_{n\to\infty} A_n = \infty$ a.s. We ask

15.) How rapidly converges $A_n$ to infinity?

Clearly in case $d \geq 3$ the volume of the largest completely covered cube around the origin does not go to infinity. However there will be a completely covered large cube somewhere.

16.) What is the volume of the largest completely covered cube in case $d \geqq 2$?

17.) Where is the largest completely covered cube located?

a) in case d=2 we ask whether the center of this cube converges to infinity

b) in case $d \geqq 3$ it is clear that the center is going to infinity but the speed is not clear.

Instead of the largest completely covered cube we can consider the largest "essentially" covered one. For example one can consider the largest integer $B_n=B_n(p)$ $(0<p<1)$ for which 100p% of the cube $[-B_n,B_n]^2$ is covered during the first n steps.

18.) Question 15-17 should be reformulated for essentially covered cubs.

Question 15 is already formulated in Erdös-Taylor (1960) where an intutive solution is also given.

The following questions look connected to the above ones

19.) Between n and $n+t_n$ $(t_n \uparrow \infty)$ how many new points will be covered? $S_j (n \leqq j \leqq t_n)$ can be considered as a newly covered point if

(i) $S_j \neq S_k$ $(k=0,1,2,\ldots,j-1)$

or

(ii) $S_j \neq S_k$ $(k=n,n+1,\ldots,j-1)$.

20.) How long time do we have to wait after n steps to obtain a new point? In fact let $Z_n$ be the smallest integer for which

$$S_{n+Z_n} \neq S_i \qquad (i=1,2,\ldots,n).$$

How can we characterize $Z_n$?

References:

Bass, R.F.-Griffin, P.S.(1985) 'The most visited site of Brownian motion and simple random walk'. Z. Wahrscheinlichkeitstheorie verw. Gebiete 70, 417-436.

Chung, K.L.-Erdös, P.(1951) 'Probability limit theorems assuming only the first moment I'. Four papers on probability. Mem. Amer. Math. Soc. No. 6.

Deheuvels, P.(1985) 'On the Erdös-Rényi theorem for random fields and sequences and its relationships with the theory of runs and spacings'. Z. Wahrscheinlichkeitstheorie verw. Gebiete 70, 91-115.

Deheuvels, P.-Devroye, L.-Lynch, J. (1986) 'Exact convergence rate in the limit theorems of Erdös-Rényi and Shepp'. Ann. Probability 14, 209-223.

Deheuvels, P.-Steinebach, J. (1986) 'Exact convergence rates in strong approximation laws for large increments of partial sums'. Preprint

Dvoretzky, A.-Erdös, P. (1950) 'Some problems on random walk in space'. Proc. Second Berkeley Symposium 353-368.

Erdös, P.-Révész, P. (1976) 'On the length of the longest head run'. Coll. Math. Soc. J.Bolyai 16, Topics in Information Theory, North Holland.

Erdös, P.-Révész, P. (1984) 'On the favourite points of a random walk'. Mathematical structures-Computational mathematics-Mathematical Modelling, 2. Sofia.

Erdös, P.-Révész, P.(1986) 'Many heads in a short block'. This volume.

Erdös, P.-Taylor, S.J. (1960) 'Some problems concerning the structure of random walk paths'. Acta Math. Acad. Sci. Hung. 11, 137-162.

Erdös, P.-Taylor, S.J. (1960) 'Some intersection properties of random walk paths'. Acta Math. Sci. Hung. 11, 231-248.

Guibas, L.J.-Odlyzko, A.M. (1980) 'Long repetitive patterns in random sequence'. Z. Wahrscheinlichkeitstheorie verw. Gebiete 53, 241-262.

Major, P. (1986) 'On the set visited once by a random walk'. Preprint.

Ortega, J.-Wschebor, M. (1984) 'On the increments of the Wiener process' Z. Wahrscheinlichkeitstheorie verw. Gebiete 65, 329-339.

Révész, P. (1982) 'On the increments of Wiener and related processes'. Ann. Probability 10, 613-622.

Samarova, S.S. (1981) 'On the length of the longest head-run for a Markov chain with two states'. Theory Probab. Appl. 26, 489-509.

# ON THE MAXIMUM LIKELIHOOD METHOD FOR CENSORED BIVARIATE SAMPLES

E. Z. Ferenstein
Institute of Mathematics
Technical University of Warsaw
Pl. Jedności Robotniczej 1
00-661 Warsaw, Poland

ABSTRACT. The result presented in this note concerns the limiting behavior of maximum likelihood estimators for type II censored samples from a bivariate absolutely continuous distribution. Let $(X_1,Y_1),\dots,(X_n,Y_n)$ be independent and identically distributed (iid) bivariate random variables (r.v's) from the parent population. One assumes that only r ($r<n$ and $r/n \to p \in (0,1)$, $n\to\infty$) pairs of (X,Y)'s are observed in which r smallest order statistics of X's occur. In these pairs suitable Y-covariates are known as induced order statistics or concomitants of order statistics. Sufficient conditions for the strong consistency and asymptotic normality of maximum likelihood estimators of a vector-valued parameter are established. The treatment is similar as in the classic iid case. Obtained results follow the asymptotics of functions of order statistics and some properties of the induced order statistics.

## 1. INTRODUCTION

Let $(X_1,Y_1)$, $(X_2,Y_2),\dots$ be a sequence of independent bivariate random variables each distributed as (X,Y). Let $X_{nj}$ be the j-th order statistics corresponding to $X_1,\dots,X_n$ and $Y_{nj}=Y_i$ if $X_{nj}=X_i$; $i,j=1,\dots,n$. $Y_{nj}$ is called the j-th induced order statistics (Bhattacharya (1974)) or the concomitant of the j-th order statistics (David (1973)).

We will consider a type II censoring experiment with respect to X's, i.e. from the sample $(X_1,Y_1),\dots,(X_n,Y_n)$ only $(X_{nj},Y_{nj})$, $j=1,\dots,r_n$, $r_n<n$, are observed. This model arises in life-testing problems in which X is the failure time of a unit selected at random and Y is a covariate whose influence on X is under investigation. Moreover, from a sample of n units only $r_n$ units with the smallest life-



*P. Bauer et al. (eds.), Mathematical Statistics and Probability Theory, Vol. B, 67–74.*

times $X_{n1} < \ldots < X_{nr_n}$ and their covariates $Y_{n1},\ldots,Y_{nr_n}$ are observed.

General distribution theory of the induced order statistics, for an arbitrary absolutely continuous bivariate distribution, has been studied by Yang (1977). Various small sample and large sample properties of the induced order statistics have been investigated by many authors, e.g. Watterson (1959), David (1973), Bhattacharya (1974), David and Galambos (1974), Sen (1981). A survey of these properties as well as their applications to the regression analysis and life-testing problems are given by Bhattacharya (1984).

For a normal distribution of (X,Y), Harrell and Sen (1979) have derived the maximum likelihood estimators of the parameters, their large sample covariance matrix, and the likelihood ratio test for independence.

In this note we are concerned with asymptotic properties of the maximum likelihood method. The strong consistency and asymptotic normality of the maximum likelihood estimators of a vector-valued parameter are established under suitable regularity conditions and the assumption that $r_n/n \to p$ as $n\to\infty$ sufficiently fast, $p \in (0,1)$. Our investigation follows the asymptotic normality of functions of order statistics (e.g. Shorack (1972), Gardiner and Sen (1978)) and the conditional independence of $Y_{nj}$, $j=1,\ldots,n$, given $X_1,\ldots,X_n$ (Bhattacharya (1974)).

In Section 2 we introduce the notation and formulate the assumptions under which the standard properties of the maximum likelihood method are preserved. Relevant theorem is stated at the end of Section 2. The proof of the theorem is given in Section 3.

## 2. ASYMPTOTIC PROPERTIES OF MAXIMUM LIKELIHOOD ESTIMATORS

Let (X,Y) have a continuous probability density function (pdf) $f_\theta(x,y)$ and distribution function (df) $F_\theta(x,y)$, $(x,y) \in R^2$, $\theta \in \Theta \subset R^k$. Let $g_\theta(x)$, $G_\theta(x)$ be the marginal pdf and marginal df of X, respectively, and let $f_\theta(y/x)$, $F_\theta(y/x)$ be the conditional pdf and conditional df of Y given X=x, respectively.

By Lemma 1 of Bhattacharya (1974); $Y_{n1},\ldots,Y_{nn}$ are conditionally independent given $X_1,\ldots,X_n$ and the conditional df of $Y_{nj}$ given $X_1=x_1,\ldots,X_n=x_n$ is $F_\theta(y/x_{nj})$. Combining this result with the joint distribution of order statistics of X's we find the joint pdf of the censored sample $(X_{n1},Y_{n1}),\ldots,(X_{nr_n},Y_{nr_n})$ as

$$n^{[r_n]} \prod_{j=1}^{r_n} f_\theta(x_j, y_j)(\overline{G}_\theta(x_{r_n}))^{n-r_n}, \quad (1)$$

where $n^{[r]} = n \cdot \ldots \cdot (n-r+1)$, $\overline{G}_\theta(x) = 1 - G_\theta(x)$, $x_1 \leq \ldots \leq x_{r_n}$.

Then, the log-likelihood function corresponding to (1) is

$$\log(n^{[r_n]}) + \sum_{j=1}^{r_n} \log f_\mu(X_{nj}, Y_{nj}) + (n-r_n)\overline{G}_\mu(X_{nr_n}), \quad (2)$$

where $\mu \in \Theta$.

For a nonnegative function $\phi_\theta(z)$, $\theta \in \Theta$, $z \in R^1$ (or $R^2$), we use the notation $\dot{\phi}_\theta(z) = (\partial/\partial\theta)\log \phi_\theta(z)$, $\ddot{\phi}_\theta(z) = (\partial/\partial\theta)\dot{\phi}_\theta(z)$, i.e. $\dot{\phi}_\theta(z)$ ($\ddot{\phi}_\theta(z)$) is the $k \times 1$ ($k \times k$) matrix of the first (second) partial derivatives of $\log \phi_\theta(z)$ with respect to $\theta$. If $A$ is a matrix, then $A^T$ stands for the transposition of $A$. If $B$ is a matrix of real valued functions, then any integral of $B$ is the matrix of appropriate integrals of corresponding elements of $B$.

Now, the likelihood equation is

$$\sum_{j=1}^{r_n} \dot{f}_\mu(X_{nj}, Y_{nj}) + (n-r_n)\dot{\overline{G}}_\mu(X_{nr_n}) = \underline{0}, \quad (3)$$

where $\underline{0} = [0, \ldots, 0]^T$.

In what follows, we assume that $\theta$ is an open set and the following regularity conditions hold:

A1. For each $\theta \in \Theta$, the partial derivatives, with respect to $\theta$, of $\log f_\theta(x,y)$ and $\log \overline{G}_\theta(x)$ exist for all $(x,y)$ and all $x$, up to the third order, inclusive. Moreover, the elements of $\dot{\overline{G}}_\theta(x)$, $\ddot{\overline{G}}_\theta(x)$ are continuous functions of $x$.

A2. For each $\theta_o \in \Theta$, there exist functions $h_i(x,y)$, $\gamma_i(x)$, $H_i(x)$; $i=1,2$; such that in a neighborhood $\theta_o$ the following relations hold:

(a) $|(\partial/\partial\theta_j) f_\theta(y/x)| < h_1(x,y)$, $|(\partial^2/\partial\theta_j\partial\theta_l) f_\theta(y/x)| < h_2(x,y)$,

where $j,l = 1,\ldots,k$ and $\int_{-\infty}^{\infty} h_i(x,y)\,dy < \infty$; $i=1,2$.

(b) $|(\partial/\partial\theta_j) g_\theta(x)| < \gamma_1(x)$, $|(\partial^2/\partial\theta_j\partial\theta_l) g_\theta(x)| < \gamma_2(x)$, where $j,l=1,\dots,k$, $\int_{-\infty}^{\infty} \gamma_i(x)\,dx < \infty$.

(c) Third partial derivatives, with respect to $\theta$, of $\log f_\theta(x,y)$ and $\log \overline{G}_\theta(x)$ are bounded in absolute values by functions $H_1(x,y)$ and $H_2(x)$, respectively, and $E_\theta H_1(X,Y) < \infty$, $H_2$ is a bounded function.

A3. There exists the matrix $E_\theta\{\dot{f}_\theta(Y/x)[\dot{f}_\theta(Y/x)]^T/X=x\}$ and it is positively definite.

Conditions A1, A2(a) and A3 insure that there exist the conditional means of $\dot{f}_\theta(X,Y)$, $\ddot{f}_\theta(X,Y)$ and the conditional covariance matrix of $\dot{f}_\theta(X,Y)$, given $X=x$, and

$$E_\theta(\dot{f}_\theta(X,Y)/X=x) = \dot{g}_\theta(x) \triangleq m_{\theta,1}(x),$$

$$\mathrm{Cov}(\dot{f}_\theta(X,Y)/X=x) = E_\theta\{\dot{f}_\theta(Y/x)[\dot{f}_\theta(Y/x)]^T/X=x\} \triangleq V_{\theta,1}(x),$$

$$E_\theta(\ddot{f}_\theta(X,Y)/X=x) = -V_{\theta,1}(x) + \ddot{g}_\theta(x).$$

In the sequel, we will need the following assumptions, satisfied for each $\theta \in \Theta$ and given $p \in (0,1)$:

B1. $E_\theta V_{\theta,1}(X)$ exists

B2. $$\lim_{c\to\infty} \sup_{x\in R^1} E_\theta[\|\dot{f}_\theta(X,Y) - m_{\theta,1}(x)\|^2 \times$$
$$\times\ I(\|\dot{f}_\theta(X,Y) - m_{\theta,1}(X)\| > c) / X=x] = 0,$$

where $\|u\|$ and $I(A)$ stand for the Euclidean norm of a vector $u$ and indicator of an event $A$, respectively.

B3. (a) $m_{\theta,1}$ is of bounded variation,

(b) $m'_{\theta,1}(x)$ and $\dot{\overline{G}}'_\theta(x)$ exist, $x \in R^1$, and are continuous at $G_\theta^{-1}(p)$.

B4. $g'_\theta(x)$ exist at $G_\theta^{-1}(p)$ and $g_\theta[G_\theta^{-1}(p)] > 0$.

B5. $\sqrt{n}(r_n/n-p) \to 0$ as $n\to\infty$.

For each $\theta \in \Theta$, let us define the matrices:

$$D_1(\theta) = \int_{-\infty}^{G_\theta^{-1}(p)} V_{\theta,1}(x)\,dG_\theta(x),$$

$$D_2(\theta) = \int_{-\infty}^{G_\theta^{-1}(p)} \int_{-\infty}^{G_\theta^{-1}(p)} (G_\theta(x) \wedge G\ (y) - G_\theta(x)G_\theta(y))\,dm_{\theta,1}(x)\,d[m_{\theta,1}(x)]^T +$$

$$+ (1-p)^2\, g_\theta^{*\prime}(p) \int_{-\infty}^{G_\theta^{-1}(p)} G_\theta(x)\,d[m_{\theta,1}(x)]^T +$$

$$+ (1-p)^2 \left( \int_{-\infty}^{G_\theta^{-1}(p)} G_\theta(x)\,dm_\theta(x) \right) [g_\theta^{*\prime}(p)]^T +$$

$$+ p(1-p)^3 g_\theta^{*\prime}(p)[g_\theta^{*\prime}(p)]^T, \text{ where } g_\theta^* = \dot{G}_\theta \circ G_\theta^{-1},$$

$$I(\theta) = D_1(\theta) + \int_{-\infty}^{G_\theta^{-1}(p)} m_{\theta,1}(x)[m_{\theta,1}(x)]^T\,dG_\theta(x) +$$

$$+ (1-p)^{-1} \int_{G_\theta^{-1}(p)}^{\infty} m_{\theta,1}(x)\,dG_\theta(x) \int_{G_\theta^{-1}(p)}^{\infty} [m_{\theta,1}(x)]^T dG_\theta(x).$$

In the sequel, the symbols $\xrightarrow{1}$, $\Rightarrow$ denote, respectively, convergence almost sure and in distribution.

Now, we may formulate the following

Theorem. If assumptions A1-A3 and B1-B6 are satisfied, then there exists a sequence $\{\hat{\theta}_n\}$ of estimators of $\theta$ such that:

(i) $P_\theta(\hat{\theta}_n$ satisfies (3)) = 1,

(ii) $\hat{\theta}_n \xrightarrow{1} \theta$ as $n \to \infty$,

(iii) $\sqrt{n}(\hat{\theta}_n - \theta) \Rightarrow N(\underline{0}, B(\theta))$, where $B(\theta) = [D_1(\theta) + D_2(\theta)]I(\theta)^{-1}$.

The proof of the theorem is given is Section 3.

Remarks. (1) In life-testing problems $G_\theta(0)=0$. Then, it is sufficient to assume that $m_{\theta,1}$ is of bounded variation on finite intervals $[0,d]$, $d>0$.

(2) If $\{r_n\}$ is a sequence of random variables such that $r_n$ are F measurable and $\sqrt{n}(r_n/n-p) \xrightarrow{1} 0$ as $n \to \infty$, then

the theorem remains true in view of Theorem 1 of Gardiner and Sen (1978).

## 3. PROOF OF THE THEOREM

First, let us consider the simple lemma which will be useful.

Lemma. Suppose the following

1. $\{(X_i,Y_i)\}$ is a sequence of iid r.v's, each distributed as the r.v. $(X,Y)$.
2. Marginal df of X is absolutely continuous.
3. h is a real valued function on $R^2$ such that $E|h(X,Y)|<\infty$.
4. $\{r_n\}$ is a sequence of naturals converging to $\infty$ so that $X_{nr_n} \xrightarrow{1} q$ as $n\to\infty$, where q is a finite constant.

Let

$$M_n=n^{-1}\sum_{j=1}^{r_n} h(X_{nj},Y_{nj}), \quad m=E[h(X,Y)I(X\le q)].$$

Then

$$M_n \xrightarrow{1} m \quad \text{as} \quad n\to\infty.$$

Proof. Let us note that $M_n=W_n+V_n$, where

$$W_n = n^{-1}\sum_{j=1}^{n} h(X_j,Y_j)I(X_j\le q),$$

$$V_n \doteq n^{-1}\sum_{j=1}^{n} h(X_j,Y_j)\{I(X_j\le X_{nr_n}) - I(X_j\le q)\}.$$

It is sufficient to show that $V_n \xrightarrow{1} 0$ as $n\to\infty$, since $W_n \xrightarrow{1} m$ as $n\to\infty$. Let $\varepsilon>0$ and $C_n$ be the event: $\{X_{nr_n}\in\Delta(\varepsilon)\}$, where $\Delta(\varepsilon)=[q-\varepsilon,q+\varepsilon]$. Denote by $D_n$ the complement of $C_n$. Now, we have

$$|V_n| < I(D_n)n^{-1}\sum_{j=1}^{n} 2|h(X_j,Y_j)| +$$

$$+I(C_n)n^{-1}\sum_{j=1}^{n} |h(X_j,Y_j)|I(X_j\in\Delta(\varepsilon)). \tag{4}$$

The first term on the right-hand side of (4) tends to 0 as

$n \to \infty$, w. pr. 1. At the same time, the second term tends to $E[|h(X,Y)|I(X \in \Delta(\varepsilon))]$ as $n \to \infty$, w. pr. 1. This last expectation converges to 0 as $\varepsilon \to 0$, which completes the proof.

Now, let us proceed on the proof of the theorem. First, let us suppose that $\theta$ is an univariate parameter. Using standard arguments one may derive, in virtue of A1, the maximum likelihood equation (3) as

$$A_n^{(1)} + A_n^{(2)}(\mu - \theta) + B_n(\mu - \theta)^2 = 0, \tag{5}$$

where

$$A_n^{(1)} = n^{-1} \sum_j \dot{f}_\theta(X_{nj}, Y_{nj}) + (1 - r_n/n)\dot{\bar{G}}_\theta(X_{nr_n}),$$

$$A_n^{(2)} = n^{-1} \sum_j \ddot{f}_\theta(X_{nj}, Y_{nj}) + (1 - r_n/n)\ddot{\bar{G}}_\theta(X_{nr_n}),$$

$$B_n = (1/2)(\varepsilon_1 n^{-1} \sum_j H_1(X_{nj}, Y_{nj}) + \varepsilon_2(1 - r_n/n)H_2(X_{nr_n})),$$

where $\sum_j = \sum_{j=1}^{r_n}$, $|\varepsilon_1| < 1$, $|\varepsilon_2| < 1$ w.pr.1.

Let us present $A_n^{(1)}$ as $T_n + S_n$, where

$$T_n = n^{-1} \sum_j m_{\theta,1}(X_{nj}) + (1 - r_n/n)\dot{\bar{G}}_\theta(X_{nr_n}),$$

$$S_n = n^{-1} \sum_j \xi_{nj}, \quad \xi_{nj} = \dot{f}_\theta(X_{nj}, Y_{nj}) - m_{\theta,1}(X_{nj}).$$

We will show that the sequences $\{A_n^{(i)}\}$, $\{S_n\}$ $\{T_n\}$ have the following asymptotic properties:

P1. $A_n^{(1)} \xrightarrow{1} 0$ as $n \to \infty$.

P2. $A_n^{(2)} \xrightarrow{1} -I(\theta)$ as $n \to \infty$.

P3. The conditional distribution of $\sqrt{n}\, S_n$, given $F = \sigma(X_1, X_2, \ldots)$, is asymptotically normal with mean 0 and variance $D_1(\theta)$.

P4. $\sqrt{n}\, T_n \Rightarrow N(0, D_2(\theta))$.

Proof of P1-P4: Let us note that in the light of assumptions B4 and B5 we have $X_{nr_n} \xrightarrow{1} G_\theta^{-1}(p)$ as $n \to \infty$. Thus P1 and P2 are consequences of the lemma and some derivations.

Assumption B2 implies that conditionally, given $F$,

$\{\xi_{nj}\}$ satisfies the Lindeberg condition. Thus the conditional distribution of $S_n$, given F, is asymptotically normal with mean 0 and variance $v_n = n^{-1}\sum_j V_{\theta,1}(X_{nj})$. Let us note that $v_n \xrightarrow{1} D_1(\theta)$ as $n \to \infty$, in the light of the lemma. Hence P3 holds.

P4 is the consequence of assumptions B3 and B5 which allow us to apply results on asymptotic normality of functions of order statistics (e.g. Shorack (1972), Gardiner and Sen (1978)).

Properties P1-P4 and standard arguments of the maximum likelihood method complete the proof in this case. In the general multivariate parameter case it is sufficient to note that linear combinations of elements of matrix coefficients in the Taylor expansion (multivariate counterpart of (5)) have again the form of statistics considered in the univariate case.

## REFERENCES

DAVID, H.A. (1973). 'Concomitants of order statistics', *Bull. Inst. Internat. Statist.* **45**, 295-300.

DAVID, H.A. and GALAMBOS, J. (1974). 'The asymptotics theory of concomitants of order statistics', *J. Appl. Probab.* **11**, 762-770.

BHATTACHARYA, P.K. (1974). 'Convergence of sample paths of normalized sums of induced order statistics', *Ann. Statist.* **2**, 1034-1039.

BHATTACHARYA, P.K. (1984). 'Induced order statistics: theory and applications', in *Handbook of Statistics*, vol. **4**, P.R. Krishnaiah and P.K. Sen, eds., Amsterdam: North Holland.

GARDINER, J.C. and SEN, P.K. (1978). 'Asymptotic normality of a class of time-sequential statistics and applications', *Comm. Statist. A - Theory Methods* **7**, 373-388.

HARRELL, F.E. and SEN, P.K. (1979). 'Statistical inference for censored bivariate normal distributions based on induced order statistics', *Biometrika* **66**, 293-298.

SEN, P.K. (1981). 'Some invariance principles for mixed rank statistics and induced order statistics and some applications', *Comm. Statist. A - Theory Methods* **10**, 1691--1718.

SHORACK, G.R. (1972). 'Functions of order statistics', *Ann. Math. Statist.* **43**, 412-427.

WATTERSON, G.A. (1959). 'Linear estimation in censored samples from multivariate normal populations', *Ann. Math. Statist.* **30**, 814-824.

YANG, S.S. (1977). 'General distribution theory of the concomitants of order statistics', *Ann. Statist.* **5**, 99--1002.

# A CHARACTERIZATION THEOREM BASED ON TRUNCATED MOMENTS AND ITS APPLICATION TO SOME DISTRIBUTION FAMILIES

Wolfgang Glänzel
Library of the Hungarian Academy of Sciences,
Information Science and Scientometrics Research Unit,
H-1361, Budapest, P.O.Box 7,
Hungary

ABSTRACT: A new representation theorem for distributions of real-valued random variables is presented. The theorem is based on a relationship between different truncated moments of the same random variable. As an example of its application, characterization theorems for some families of both continuous and discrete distributions are derived. Further applications can be obtained after certain transformations. These characterizations may also serve as a basis for parameter estimation.

## 1. INTRODUCTION

In the recent years a great interest has been focused on characterization of distributions by truncated moments. Some important general theorems were developed by Kotz & Shanbhag (1980). Under stronger conditions a characterization theorem based on a proportional relation between different truncated moments of both continuous and discrete real random variables was proved and applied to some particular distributions (Glänzel & al. 1984, Glänzel 1986 a,b). In the followings, a more general result will be given, which shows the validity of the above mentioned characterization principle under very mild conditions.

Additionally, applications to wide classes of continuous and discrete distributions of practical significance will be presented.

## 2. THE CHARACTERIZATION THEOREM

Let $(\Omega,\mathcal{A},P)$ be a given probability space and let $X: \Omega \to H$ be a random variable, where $H = (a,b]$ for some real $a < b$; $a = -\infty$ and $b = +\infty$ are not excluded. For a given real function h defined on H we consider the function

$$E(h(X)|X\geq x) = e_h(x) \quad ; \; x \in H$$

provided it is defined. Let F be the distribution function of the random variable X. Put $G = 1 - F$.



P. Bauer et al. (eds.), Mathematical Statistics and Probability Theory, Vol. B, 75–84.

Theorem: Let g and h be two real functions defined on H such that

$$e_g = e_h \lambda \tag{2.1}$$

is defined. Assume that $g,h \in C^1(H)$, the function g/h is of bounded variation and λ is a left-continuous function with an at most countable set of discontinuity points on H. Further assume that the equation

$$\lambda h - g = 0 \tag{2.2}$$

has no solution on int H. Finally, assume that

$$\int_{[a,b]} \frac{\lambda h' - g'}{\lambda h - g}\, dx = +\infty \;. \tag{2.3}$$

Then F is uniquely determined by the functions g, h and λ.

*Proof* : We prove the theorem in two steps. First we show that the function $K(x) = \int_{[x,b]} h\, dF$ $(x \in H)$ is uniquely determined by λ, g and h.

Consider Eq. (2.1)

$$\int_{[x,b]} g\, dF = \int_{[x,b]} h \frac{g}{h}\, dF = \lambda \int_{[x,b]} h\, dF$$

and put $g(x)/h(x) = s(x)$ on H.

Hence we have

$$\lambda(x)K(x) = - \int_{[x,b]} \{ \int_{[x,t]} ds(\tau) + s(x)\}\; dK(t).$$

The application of Fubini's theorem leads to the following equation:

$$\frac{K(x)}{\int_{[x,b]} K(t)\, ds(t)} = \frac{1}{\lambda(x) - s(x)} \tag{2.4}$$

Note that $B(x) = \int_{[x,b]} K(t)$ is continuous and almost everywhere differentiable on H. That is why B(x) cannot change its sign. Hence the following manipulation is justified:

$$- \int_{[a,y]} \frac{dB(x)}{B(x)} = \int_{[a,y]} \frac{K(s)ds(x)}{\int_{[x,b]} K(t)ds(t)} = - \ln \{ \frac{B(y)}{b(a)}\} \;.$$

Hence and from Eq. (2.4) it follows that:

$$K(y) = K(a) \frac{\lambda(a)-s(a)}{[\lambda(y)-s(y)]} \exp \{- \int_{[a,y]} \frac{ds(x)}{\lambda(x)-s(x)} \}. \quad (2.5)$$

Thus we have completed step one. Eq. (2.5) can be written as follows:

$$K(y) = K(a) \frac{\lambda(a)h(a)-g(a)}{\lambda(y)h(y)-g(y)} \exp \{ \int_{[a,y]} \frac{\lambda(t)h'-g'(t)}{\lambda(t)h(t)-g(t)} dt\} \quad (2.6)$$

Consider the following identity:

$$G(y) = - \int_{[y,b]} \frac{1}{h} dK = - \int_{[y,b]} \{ \int_{[y,t]} d(\frac{1}{h}) + \frac{1}{h(y)}\} dK(t)$$

Hence we obtain by application of Fubini's theorem:

$$G(y) = \frac{K(y)}{h(y)} - \int_{[y,b]} \frac{h'(t)}{\{h(t)\}^2} K(t) dt .$$

The substitution of Eq. (2.6) into the last equation leads directly to the statement. Note that 1-G is a distribution only if condition (2.3) holds. This completes the proof.

*Remark 2.1.* If $h \equiv 1$, then the distribution function F is uniquely determined by the truncated moment function $e_g$ (Kotz & Shanbhag 1980).

*Remark 2.2.* Let g, h and $\lambda$ be real functions satisfying the conditions of the theorem. Let $g^*$ and $h^*$ be real functions defined on H such that

$$h^* = \alpha h + \beta \ ; \quad \alpha \in \mathbb{R}\setminus\{0\} \quad \beta \in \mathbb{R}$$

and

$$g^* = \gamma g + \delta; \quad \gamma \in \mathbb{R}\setminus\{0\} \ , \ \delta \in \mathbb{R} .$$

Then Eq. (2.1) is equivalent to

$$(2.7) \qquad e_{g*} = \psi . e_{h*} + \phi \qquad (2.7)$$

where $\psi = (\gamma/\alpha).\lambda$ and $\phi = -\beta\psi+\delta$ .

<u>Corollary 2.1.</u> Assume that the conditions of the theorem are satisfied. Assume further that the function $\lambda$ is continuously differentiable on H. Then the distribution function F is strictly monotone and continuously differentiable, and for its density we have

$$f(x) = C \frac{\lambda'}{\lambda h-g} \exp \left\{ - \int_a^x \frac{\lambda' h}{\lambda h-g} dt \right\} , \quad (2.8)$$

where the constant C has to be chosen so that $\int_a^b f \, dx = 1$.

Corollary 2.2. Assume that the conditions of the theorem are satisfied. Assume that X is an integer valued random variable. Then the distribution $P(X=k) = p_k$ $(k \in H)$ is uniquely determined by g, h and λ, particularly

$$p_k = \begin{cases} C\,\dfrac{\lambda(k+1)-\lambda(k)}{\lambda(a+1)-\lambda(a)} \displaystyle\prod_{i=a}^{k-1} \dfrac{\lambda(i)h(i)-g(i)}{\lambda(i+2)h(i+1)-g(i+1)} & \text{if } b=\infty \text{ or } b<\infty \text{ and } k<b \\ C\,\dfrac{\lambda(b-1)h(b-1)-g(b-1)}{\{\lambda(a+1)-\lambda(a)\}h(k)} \displaystyle\prod_{i=a}^{b-2} \dfrac{\lambda(i)h(i)-g(i)}{\lambda(i+2)h(i+1)-g(i+1)} & \text{if } b<\infty \text{ and } k=b, \end{cases} \tag{2.9}$$

where the constant C is found from the condition $\sum_H p_k = 1$.

## 3. APPLICATIONS

### 1. The Distributions of Irwin's System

Definition 3.1. We say that the distribution of a given nonnegative integer valued random variable X belongs to Irwin's system (Irwin 1975) if

$$p_{k+1} = \frac{(k+\beta)(k+\gamma)}{(k+\alpha+\beta+\gamma)(k+1)}\, p_k \; ; \; k \geq 0, \tag{3.1}$$

where α, β and γ are parameters such that $p_k$ are probabilities.

Excluding the case of conjugate complex parameters β and γ, all distributions of Irwin's system occur in the classical Pólya-Eggenberger urn model.

The following proposition was proven by Glänzel & et al. (1984).

Proposition 3.1. Let X be a nonnegative integer valued random variable. Then X has a distribution belonging to Irwin's system, iff the functions in Eq. (2.7) have the forms

$$\phi(k) = ak + b; \quad a > 0,\ b > 0$$

and

$$\psi(k) = ck; \quad c + a_o b + 2a - 1 > 0 \quad (a_o = \min(2-1/a, 1))$$

for $g^*(x)=x$ and $h^*(x)=1/(1+x)$ $(x \geq 0)$, provided this defines a distribution.

The connection between the parameters a, b and c and the parameters α, β and γ in Eq. (3.1) is as follows:

$$a = \frac{\alpha}{\alpha-1}, \quad b = \frac{\beta\gamma}{\alpha-1}, \quad c = -\frac{(\beta-1)(\gamma-1)}{\alpha-1}.$$

Consider the following cases resulting from special choices of the parameters a, b and c (cp. Glänzel & al. 1984).

Case 1.1. $a > 1$

a) $-2a-b+1<c\leq-a-b+1-2\sqrt{b(a-1)}$: Hypergeometric distribution. This is possible only if $a>\sqrt{b(a-1)}$. Note that the smaller root of the quadratic expression $(a-1)k^2+(a+b+c-1)k+b$ has to be a natural number. Extensions for real roots were given by Kemp & Kemp (1956).

b) $\max\{-2a-b+1,-a-b+1-2\sqrt{b(a-1)}\}<c<-a-b+1+2\sqrt{b(a-1)}$ : This corresponds to the case when the parameters $\beta$ and $\gamma$ are complex conjugate (Irwin 1975).

c) $-a-b+1+2\sqrt{b(a-1)}\leq c$; Inverse Pólya-Eggenberger distribution. The special case $c=0$ corresponds to a Waring distribution.

Case 1.2. $a = 1$

a) $-b-1<c<-b$: Binomial distribution if $b/(b+c)$ is an integer. The parameters are $n = -b/(b+c)$ and $p = -(b+c)$.

b) $c=-b$: Poisson distribution with the parameter $b$.

c) $c>-b$: Negative binomial distribution. The parameters are $N = b/(b+c)$ and $q = (b+c+1)^{-1}$.

Case 1.3. $a < 1$

a) $c>-2a-(2-1/a)b+1$: Pólya-Eggenberger (negative hypergeometric) distribution. Note that for $a<0.5$ $c$ has to be positive.

b) $c=0$ and $a=0.5$: Discrete uniform distribution if $2b \in \mathbb{N}$ .

Fig. 1 shows the domains of Irwin's system on the $(a,c)$ parameter plane at fixed $b$ ($b=1$) (cp. Schubert & al. 1985).

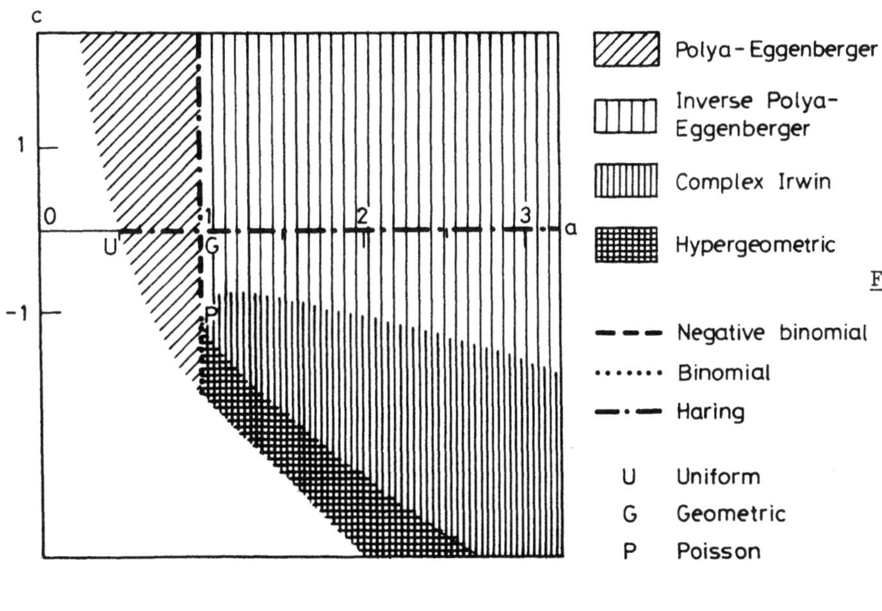


Fig. 1.

## 2. The Pearson System

Definition 3.2. The distribution of a continuous random variable $X:\Omega\to H$ belongs to the Pearson system, if the density function f is differentiable and satisfies the following equation:

$$\frac{d \log f}{dx}(x) = -\frac{x + A}{Bx^2 + Cx + D} \quad ; \; x \in H, \tag{3.2}$$

where A, B, C and D are real parameters such that f is a probability density.

Proposition 3.2. Let $X:\Omega\to H$ be a continuous random variable and let $g^*(x) = x^2$ and $h^*(x) = x$. Assume that $e_{g^*}$ and $e_{h^*}$ exist and are differentiable on H. Then the distribution of the random variable X belongs to the Pearson system if the functions in Eq. (2.7) have the form

$$\psi(x) = r\cdot x + t \quad \text{and} \quad \phi(x) = u\cdot x + w ,$$

where r, t, u and w are some real parameters such that the distribution is well-defined. The reverse statement holds if $(Bx^3+Cx^2+Dx)f(x) \to 0$ as $x \to b$.

*Proof:* Assume that the truncated moments in question exist and consider the following particular form of Eq. (2.7)

$$E(X^2|X\geq x) = E(X|X\geq x)(rx+t) + ux+w \; ; \; x \in H . \tag{3.3}$$

The differentiability of both sides of Eq. (3.3) is evident. Since $\psi$ and $\phi$ are differentiable, the density function f is differentiable, too. From Eq. (3.3) we obtain by repeated differentiation

$$\frac{d \log f}{dx}(x) = -\frac{(3r-2)x + (t+2u)}{(r-1)x^2 + (t+u)x + w} \quad ; \quad x \in H,$$

i.e., the distribution of X is a member of the Pearson system.

Assume now that X has a Pearson-type distribution. Then Eq. (3.2) is satisfied for all $x \in H$ and we have

$$-(x + A)\cdot f = (Bx^2 + Cx + D)\cdot f' \; ; \quad x \in H .$$

Hence

$$-E(X|X\geq y)-A=-2B\cdot E(X|X\geq y)-C + (By^2+Cy+D)\cdot f(y|X\geq y) \; ; \; y \in H. \tag{3.4}$$

In an analogous manner we obtain from

$$-(x^2+Ax)\cdot f = (Bx^3 + Cx^2 + Dx)\cdot f'; \; x \in H.$$

The following equation:

$$-E(X^2|X\geq y)-A\cdot E(X|X\geq y) = -3B\cdot E(X^2|X\geq y)-2C\cdot E(X|X\geq y)-D + \\ + y(By^2+Cy+D)\cdot f(y|X\geq y); \; y \in H. \tag{3.5}$$

Eqs (3.4) and (3.5) lead directly to (3.3) with $r = (2B-1)/(3B-1)$,

$t = (A-2C)/(3B-1)$, $u = (C-A)/(3B-1)$ and $w = -D/(3B-1)$. Thus the proof is completed.

*Remark 3.1.* Note that r has to be positive for all Pearson-type distributions.

Since it can always be assumed that the expectation of a continuous random variable is zero, in the following short discussion u=0 and w>0.

Case 2.1. $0 < r < 1$

a) $t \neq 0$ Pearson Type I. Finite beta distribution. The density function is

$$f(x) = \frac{1}{B(p,q)} \frac{(x-a)^{p-1}(b-x)^{q-1}}{(b-a)^{p+q-1}}; \; x \in (a,b).$$

The parameters of the finite beta distribution are:

$$a = \frac{-t + \{t^2-4w(r-1)\}^{1/2}}{2(r-1)} < 0 ,$$

$$b = \frac{-t - \{t^2-4w(r-1)\}^{1/2}}{2(r-1)} > 0 ,$$

$$p = \frac{-rt - r\{t^2-4w(r-1)\}^{1/2}}{2(r-1)\{t^2-4w(r-1)\}^{1/2}} > 0,$$

$$q = \frac{rt - r\{t^2-4w(r-1)\}^{1/2}}{2(r-1)\{t^2-4w(r-1)\}^{1/2}} > 0 .$$

b) $t = 0$ Pearson Type II. This is a symmetric beta distribution with the parameters

$-r/(2r-2) = p = q > 0$ and $\sqrt{w/(1-r)} = b = -a > 0,$

If $r = 2/3$, a uniform distribution is obtained.

Case 2.2. $r = 1$

a) $t \neq 0$ Pearson Type III. This is the family of gamma distributions. The general form of their densities is

$$f(x) = \frac{(x-\gamma)^{\alpha-1} \exp\{-(x-\gamma)/\beta\}}{\beta^{\alpha}\Gamma(\alpha)} \; ; \; x \in H$$

where $\gamma = -w/t$, $\alpha = w/t^2$, $\beta = t$ and

$$H = \begin{cases} (-w/t,\infty), & \text{if } t > 0 \\ (-\infty,-w/t), & \text{if } t < 0 . \end{cases}$$

b) $t = 0$ normal distribution.

Case 2.3. $1 < r < 1 + t^2/4w \quad (t \neq 0)$

a) Pearson Type VI. Infinite beta distribution. The density function is

$$f(x) = \frac{1}{B(-p-q+1,q)} \cdot \frac{(x-a)^{p-1}(x-b)^{q-1}}{(b-a)^{p+q-1}} \; ; \; x > b.$$

The parameters are defined in Case 2.1. $E(X) = 0$ implies the condition $0 > b > a$, i.e., $t$ must be positive. The condition $-(p-1) > q-1 > -1$ is obviously satisfied.

Case 2.4. $r = 1 + t^2/4w \quad (t \neq 0)$. Pearson Type V.

$$f(x) = \frac{\beta^{\alpha}\exp\{-\beta/(x-\gamma)\}}{\Gamma(\alpha)\;(x-\gamma)^{\alpha-1}} \; ; \; x \in H,$$

where $\gamma = -2w/t$, $\alpha = -2(t^2+2w)/t^2$, $\beta = 2w(t^2+4w)/t^3$

$$\text{and} \quad H = \begin{cases} (-2w/t,\infty), & \text{if } t > 0 \\ (-\infty,-2w/t), & \text{if } t < 0 . \end{cases}$$

Pearson-type V distribution can be considered as "inverse gamma distributions".

Case 2.5. $r > 1 + t^2/4w$

a) $t \neq 0$ Pearson Type VI. The density function is

$$f(x)=K\cdot\{c^2+(x-\gamma)^2\}^{-m}.\exp\{-\beta\tan^{-1}((x-\gamma)/c)\}; \quad x \in \mathbb{R},$$

where K is a constant chosen so that $\int_H f(t)dt = 1$.

The parameters are:

$m = (3r-2)/(2r-2)$; $c = \{4w(r-1)-t^2\}^{1/2}/(2r-2)$; $\gamma = -t/(2r-2)$;

$$\beta = \frac{-rt}{\{4w(r-1)-t^2\}^{1/2}(r-1)} .$$

b) $t = 0$ Pearson Type VII. The density function derived from the preceding one is

$$f(x) = \frac{\Gamma(m)\cdot c^{2m-1}}{\pi^{1/2}\Gamma(m-1/2)(c^2+x^2)^m} \; ; \; x \in \mathbb{R},$$

where $m = (3r-2)/(2r-2)$ and $c = \{w/(r-1)\}^{1/2}$.
If $w > 1$ and $r$ ; $(1+w)/2$, a (central) t-distribution is obtained.

Fig. 2. shows the main domain of the Pearson system on the $(t,r)$ plane. $u = 0$ is assumed as above.

## 3. The Standard Cauchy Distribution

Proposition 3.3. Let $X:\Omega\to H$ be a continuous random variable. Then X

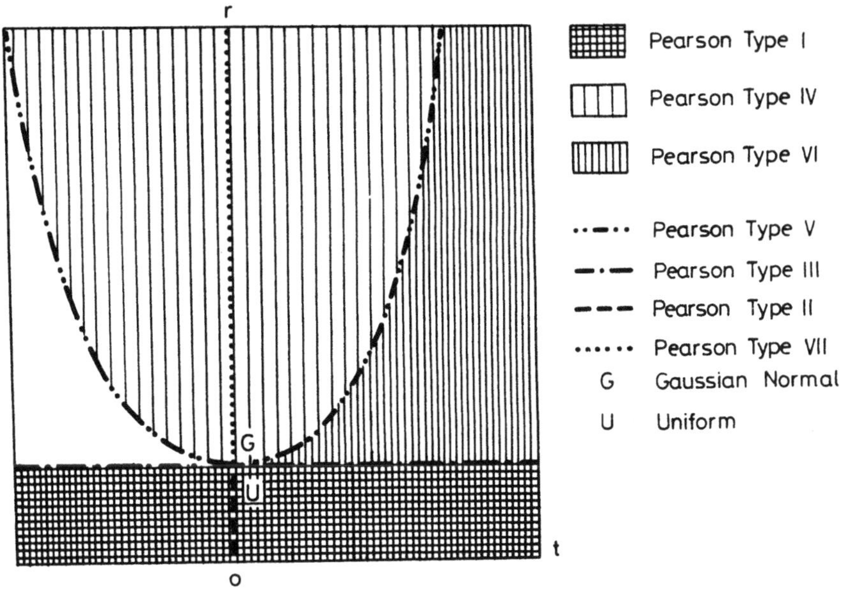


Fig. 2.

has a standard Cauchy distribution if and only if the functions in the theorem can be chosen as:

$$g(x) = (1-x^2)/(1+x^2),\ h(x) = -2x/(1+x^2) \quad \text{and } \lambda = x\ ;\ x \in \mathbb{R}\ .$$

*Proof:* The necessity is obvious. Assume that the following holds:

$$E\left(\frac{1-X^2}{1+X^2}\middle| X\geq x\right) = E\left(\frac{-2X}{1+X^2}\middle| X\geq x\right).x\ ;\quad x\ \in\mathbb{R}. \tag{3.6}$$

It is clear that both sides of Eq. (3.6) are differentiable. Since g, h and $\lambda$ are differentiable, the density f is differentiable, too. From Eq. (3.6) we get by repeated differentiation

$$\frac{1}{f(x)}\ \frac{df}{dx}\ (x) = -\ \frac{2x}{1+x^2}$$

i.e., under the given conditions the random variable X has a standard Cauchy distribution.

*Remark 3.2.* Let z be a continuously differentiable and strictly monotone real function. Then the above applications can be extended using the transformation z(X) since this transformation does not change the conditions of the theorem. Thus a characterization of Cauchy distributions with any location and scale parameters can be directly derived from the characterization of the standard Cauchy distribution using a linear transformation.

## REFERENCES

Glänzel, W., Telcs, A., Schubert, A.: Characterization by Truncated Moments and Its Application to Pearson-type Distributions. Z. Wahrscheinlichkeitstheorie verw. Gebiete, 66 (1984) 173-183.

Glänzel, W.: A Characterization of the Normal Distribution. Studie Sci. Math. Hungarica (to be published in 1987).

Glänzel, W.: A Characterization Theorem for Pearson-Type Distributions. (Submitted to Sankhyā Ser. A.).

Irwin, J.O.: The Generalized Waring Distribution; Parts I, II, III. J.R. Statist. Soc. B, 18 (1975) 202-211.

Kotz, S., Shanbhag, D.N.: Some New Approaches to Probability Distributions. Adv. Appl. Probability, 12 (1980) 903-921.

Schubert, A., Glänzel, W., Mérő, A.: Classification and Identification of Frequency Distributions Using a Characterization Theorem. In: Sváb, J., Győrffy, M., Ábrányi, A., Kovács, G. (Eds.), Proc. of the First European Biometric Conference of the Biometric Society, held in Budapest, on April 1-3, 1985.

A CLASS OF NON-PARAMETRICALLY CONSTRUCTED PARAMETER ESTIMATORS FOR A STATIONARY AUTOREGRESSIVE MODEL

W. González Manteiga and J.M. Vilar Fernández

Santiago de Compostela

ABSTRACT.

This article presents a new class of estimators for the parameters $\theta^t = (\theta_1, \dots, \theta_q)$ of the stationary autoregressive model

AR(q), $X_t = \sum_{i=1}^{q} \theta_i X_{t-i} + \varepsilon_t$, with $E|X_t| = 0$, $E[\varepsilon_t] = 0$ and $Var(\varepsilon_t) = \sigma^2$. The new estimators are obtained by minimizing the functional

$$\hat{\psi}(\theta) = \int (\hat{\alpha}_n(\vec{x}) - \vec{x}^t\theta)^2 \hat{f}_n(\vec{x})\, d\vec{x}$$

where $\hat{\alpha}_n$ and $\hat{f}_n$ are respectively non-parametric estimators of the prediction function

$\alpha(\vec{x}) = \alpha(x_1, \dots, x_q) = E[X_t / X_{t-1} = x_1, \dots, X_{t-q} = x_q] = \theta^t\vec{x}$ and

of $f$, the stationary initial density of $(X_1, \dots, X_q)$. Consistency and asymptotic normality properties are proved.

Let $\{X_t\}_{t \in Z^+}$ be a stationary time series of the type AR(q)

$$X_k = \theta_1^0 X_{k-1} + \dots + \theta_q^0 X_{k-q} + \varepsilon_k = (X_{k-1}, \dots, X_{k-q}) \begin{pmatrix} \theta_1^0 \\ \vdots \\ \theta_q^0 \end{pmatrix} + \varepsilon_k =$$

$$= \vec{X}_{k-1}^t \theta_0 + \varepsilon_k \tag{1.1}$$



*P. Bauer et al. (eds.), Mathematical Statistics and Probability Theory, Vol. B, 85–95.*

where k-q>0 and for convenience $E[X_t] = 0$.

It will be assumed that the stationary initial distribution on $\mathbb{R}^q$ of $\vec{X}_q$ has a density f with respect to Lebesgue measure on $\mathbb{R}^q$ and the error process $\{\varepsilon_t\}$ has also a density on $\mathbb{R}$. Denoting by $F_q$ and F the d.f.s then $F_{q+1}$ will be the joint distribution of $(\vec{X}_q, X_{q+1})$ on $\mathbb{R}^{q+1}$ with respect to $F_q$ and the distribution F.

With the supposition of that $\{\varepsilon_t\}$ is white noise, i.e. independent identically distributed random variables with mean 0 and variance $\sigma^2$, the vector $\theta_0$ is the value of $\theta$ which minimizes

$$\psi(\theta) = E[(X_k - \vec{X}^t_{k-1}\theta)^2] = \int (y - \vec{x}^t\theta)\, dF_{q+1}(\vec{x}, y) \tag{1.2}$$

The traditional way of estimating $\theta_0$ from a sample $\{X_1, \ldots, X_n\}$, n>q obeying (1.1), the least squares method, consists in minimizing

$$\psi^*(\theta) = \int (y - \vec{x}^t\theta)^2 dF^n_{q+1}(\vec{x}, y) = \frac{1}{(n-q)} \sum_{i=q+1}^{n} (X_i - \vec{X}^t_{i-1}\theta)^2 \tag{1.3}$$

where $F^n_{q+1}$ is the empirical distribution associated with $\{(\vec{X}_q, X_{q+1}), \ldots, (\vec{X}_{n-1}, X_n)\}$. In other words, the least squares method minimizes the empirical estimate of the functional (1.2).

The class of estimators presented in this paper is based in different methods of construction, defining estimators $\hat{\theta}_n$ obtained by subjecting the sample data to smmothing techniques (Titterington, 1985). To be precise, given a nonparametric estimator $\hat{\alpha}_n$ of the prediction function

$\alpha(\vec{x}) = \alpha(x_1, \ldots, x_q) = E[X_{q+1} / \vec{X}_q = \vec{x}]$, the estimator of $\theta_0$ is defined as the value of $\theta$ which minimizes the functional

$$\hat{\psi}(\theta) = \int (\hat{\alpha}_n(\vec{x}) - \vec{x}^t\theta)^2 d\Omega_n(\vec{x}) \tag{1.4}$$

where $\Omega_n$ is a weighting function likewise constructed from the sample. These new estimators are a generalization (for dependent data) of the "smooth regression parameter estimators" which we have described previously for general linear regression models (Faraldo-Roca and González-Manteiga, 1985; Cristóbal et al., 1986), and might similarly be referred to as "smooth autoregression parameter estimators".

Is interesting observe that the least squares estimators defined in (1.3) represents and extreme special case in (1.4) when

$\hat{\alpha}_n(\vec{x}) = \sum_{i=q+1}^{n} X_i\, I_{\{\vec{X}_{i-1}\}}(\vec{x})$ and $\Omega_n(\vec{x}) = F^n_q(\vec{x})$ the empirical distribution function constructed with $\{\vec{X}_q, \ldots, \vec{X}_{n-1}\}$.

Details of recent advances in non-parametric estimation from dependent data may be found in Collomb (1982), Masry (1983), Bierens (1983), Collomb and Doukhan (1983), Bosq (1983), Hart (1984), Yakowitz (1985), Collomb and Härdle (1986) and other articles.

How almost all non-parametric prediction function estimators have the form

$$\hat{\alpha}_n(\vec{x}) = \Big(\sum_{i=q+1}^{n} X_i \delta_m(\vec{x},\vec{X}_{i-1})\Big)\Big/\Big(\sum_{i=q+1}^{n} \delta_m(\vec{x},\vec{X}_{i-1})\Big)$$

where $\{\delta_m: \mathbb{R}^q \times \mathbb{R}^q \longrightarrow \mathbb{R}\}$ is a sequence of measurable functions and using the fact that for i= q... exists a stationary density f associated with $\vec{X}_i$, we can define the weighting function

$$\Omega_n(\vec{x}) = \int_{-\infty}^{\vec{x}} \hat{f}_n(\vec{t})\,d\vec{t} = \frac{1}{(n-q)} \sum_{i=q+1}^{n} \int_{-\infty}^{\vec{x}} \delta_m(\vec{t},\vec{X}_{i-1})\,d\vec{t}$$

where $\hat{f}_n(\vec{x}) = (\sum_{i=q+1}^{n} \delta_m(\vec{x},\vec{X}_{i-1}))/(n-q)$ is a non-parametric estimate of the density (the observation $X_n$ has been omitted to simplify the notation). The expression for the estimator $\hat{\theta}_n$ now obtained by minimizing (1.4) is

$$\hat{\theta}_n = \left(\int \vec{x}\vec{x}^t \hat{f}_n(\vec{x})\,d\vec{x}\right)^{-1} \int \hat{\alpha}_n(\vec{x}) \hat{f}_n(\vec{x})\,\vec{x}\,d\vec{x} =$$

$$= \left( \frac{\sum_{i=q+1}^{n} \int \vec{x}\vec{x}^t \delta_m(\vec{x},\vec{X}_{i-1})\,d\vec{x}}{(n-q)} \right) \left( \frac{\sum_{i=q+1}^{n} \int \vec{x}\delta_m(\vec{x},\vec{X}_{i-1}) X_i\,d\vec{x}}{(n-q)} \right) \qquad (1.5)$$

which will be of importance in connection with

$$\theta_0 = \left(\int \vec{x}\vec{x}^t dF_q(\vec{x})\right)^{-1} \int \vec{x} y\, dF_{q+1}(\vec{x},y) \qquad (1.6)$$

determined uniquely if $(E[\vec{X}_t\vec{X}_t^t])^{-1}$ exists.

An interesting special case of the class (1.5) is obtained as follows. Taking

$$\delta_m(\vec{x},\vec{u}) = \frac{1}{(\varepsilon(n))^q} \prod_{i=1}^{q} K\Big(\frac{x_i - u_i}{\varepsilon(n)}\Big) = \frac{1}{(\varepsilon(n))^q} K^*\Big(\frac{\vec{x}-\vec{u}}{\varepsilon(n)}\Big)$$

where K is a symmetric, positive onedimensional kernel satisfying

$$\int K(z)\,dz = 1, \qquad \int zK(z)\,dz = 0$$

and

$$\int z^2 K(z)\,dz < \infty$$

with $\varepsilon(n)$ the traditional window in nonparametric estimation. This yields

$$\hat{\theta}_n = \left( \sum_{i=q+1}^{n} \vec{X}_{i-1} \vec{X}^t_{i-1} + kI_q \right)^{-1} \left( \sum_{i=q+1}^{n} \vec{X}_{i-1} X_i \right)$$

where $k= n(\varepsilon(n))^2 \int z^2 K(z)dz$, i.e. a time series version of the ridge regression estimator. Details of the proof for the ordinary regression case are given in Cristóbal et al (1986).

The purpose of this article is the study of asimptotic properties of $\hat{\theta}_n$. For this reason we will use certain properties of $\{X_t\}$ and will be assumed the following properties on $\{\delta_m\}$:

A) $\delta_m > 0$, $\int \delta_m(\vec{x},\vec{u})d\vec{u} = 1$ and $\delta_m(\vec{x},\vec{u}) = \delta_m(\vec{u},\vec{x})$ for all $\vec{x} \in \mathbb{R}^q$ and $(\vec{x},\vec{u}) \in \mathbb{R}^{2q}$.

B) $h^t_m(\vec{u}) = \left( \int \delta_m(\vec{x},\vec{u}) x^k_r x^{m^*}_s d\vec{x} \right)^t \to (u^k_r u^{m^*}_s)^t = h^t(\vec{u})$

for $k,m^* = 0,1,2$ $r,s= 1,\ldots,q$ and $t= 1,2$ with $\int h^t(\vec{u}) f(\vec{u}) d\vec{u} < \infty$ and $h^t_m(\vec{u}) \leq g(\vec{u})$ for some $g(\vec{u})$ such that $\int g(\vec{u}) f(\vec{u}) d\vec{u} < \infty$.

C) $\sup_{u \in \Delta} \left| \left( \int x^k_r x^{m^*}_s \delta_m(\vec{x},\vec{u}) d\vec{x} \right)^t - (u^k_r u^{m^*}_s)^t \right| \to 0$ for $k,m^* = 0,1,2$, $r,s= 1,\ldots,q$ and $t= 1,2$ with $\Delta$ a compact subset of $\mathbb{R}^q$.

Note that conditions of the type A), B) and C) on the $\delta_m$ are satisfied for a wide class of non-parametric estimators including histogram and kernel estimators between others.

Note also that $\{X_t\}$ may be expressed in the form

$$X_t = \sum_{j=0}^{\infty} a_j \varepsilon_{t-j}$$

and using the mutual independence of $\{\varepsilon_t\}$

$$E\left[X_t^2\right] = \sigma^2 \sum_{j=0}^{\infty} a_j^2 < \infty ,$$

then in terminology of Billingsley (1968) $\{X_t\}$ is $L^2$-stable with respect to $\{\varepsilon_t\}$, since

$$\gamma(l) = E\left[(X_t - X_{t,l})^2\right] = \sigma^2 \sum_{j=l+1}^{\infty} a_j^2 \to 0, \text{ where } X_{t,l} = \sum_{j=0}^{l} a_j \varepsilon_{t-j}$$

is the $l$-$L^2$ approximation to $X_t$. Moreover if we consider

$$\begin{pmatrix} X_t \\ \vdots \\ X_{t-q} \end{pmatrix} = \begin{pmatrix} \sum_{j=0}^{\infty} a_j \varepsilon_{t-j} \\ \vdots \\ \sum_{j=0}^{\infty} a_j\ {}_{t-q-j} \end{pmatrix} = \sum_{j=0}^{\infty} a_j \begin{pmatrix} \varepsilon_{t-j} \\ \vdots \\ \varepsilon_{t-q-j} \end{pmatrix} = \sum_{j=0}^{\infty} a_j \vec{\eta}_{t-j} =$$

$$= \vec{Y}_t = \begin{pmatrix} X_t \\ \vec{X}_{t-1} \end{pmatrix},$$ then $\{\vec{\eta}_t\}$ is a $(q+1)$-dimensional time series with covariance matrices

$$J_1 = \begin{pmatrix} 0 & \sigma^2 & & & 0 \\ & \cdot & \cdot & & \\ & & \cdot & \cdot & \sigma^2 \\ & & & \cdot & \\ 0 & & & \cdot & 0 \end{pmatrix}, \dots, J_q = \begin{pmatrix} 0 & & & & \sigma^2 \\ & \cdot & & & \\ & & \cdot & & \\ & & & \cdot & \\ & & & \cdot & 0 \end{pmatrix}$$

and $\{\vec{Y}_t\}$ is an $L^2$-stable process with respect to $\{\vec{\eta}_t\}$.

THEOREM: With $\hat{\theta}_n$ defined as in (1.5),

a) If A and B hold (B uniformly over the density of all 1-approximations) then

$$\hat{\theta}_n \xrightarrow{P} \theta_0 ;$$

b) If A and C hold and the support of f, S, is compact with $E[X_t^4]<\infty$ and $\Delta= S$, then

$$\hat{\theta}_n \xrightarrow{a.s.} \theta_0 ;$$

c) with the same hypotheses as in (a), $E[X_t^4]<\infty$ and assuming further that the support of $\delta_m(\vec{x},\cdot)$ is compact, that $\sup_{\vec{u}} \delta_m(\vec{x},\vec{u}) = O(m^q)$ for all $\vec{x}\in\mathbb{R}^q$ and $\delta_m(\vec{x},\vec{u}) = 0$ if $\|\vec{x}-\vec{u}\| > c\varepsilon_n$ for all $\{\varepsilon_n\}$ such that $n\varepsilon_n^2 \to 0$, with $c\in\mathbb{R}^+$, then.

$$\sqrt{n}(\hat{\theta}_n-\theta_0) \xrightarrow{d} N_q(0,\sigma^2(E[\vec{X}_t\vec{X}_t^t])^{-1}).$$

Proof of a)

In view of (1.5) and (1.6), the proof of (a) consists in showing that

$$\sum_{i=q+1}^{n} \frac{\int x_r x_s \delta_m(\vec{x},\vec{X}_{i-1})\,d\vec{x}}{(n-q)} \xrightarrow{P} E[X_r X_s] \tag{1.7}$$

and that

$$\sum_{i=q+1}^{n} \frac{\int x_r X_i \delta_m(\vec{x},\vec{X}_{i-1})\,d\vec{x}}{(n-q)} \xrightarrow{P} E[X_r X_{q+1}] \tag{1.8}$$

where $r, s = 1, \ldots, q$ are indices of $\vec{X}_q^t = (X_1, \ldots, X_q)$.

To prove (1.7), we write

$$\frac{1}{(n-q)} \sum_{i=q+1}^{n} \{ \int x_r x_s \delta_m(\vec{x}, \vec{X}_{i-1}) d\vec{x} - E[X_r X_s] \} = \Delta_1 + \Delta_2 ,$$

where

$$\Delta_1 = \frac{\sum_{i=q+1}^{n} \{ \int x_r x_s \delta_m(\vec{x}, \vec{X}_{i-1}) d\vec{x} \}}{(n-q)} - \frac{\sum_{i=q+1}^{n} E[\int x_r x_s \delta_m(\vec{x}, \vec{X}_{i-1}) d\vec{x}]}{(n-q)} =$$

$$= \left( \sum_{i=q+1}^{n} R_i \right) / (n-q), \text{ and}$$

$$\Delta_2 = \sum_{i=q+1}^{n} \frac{E[\int x_r x_s \delta_m(\vec{x}, \vec{X}_{i-1}) d\vec{x}]}{(n-q)} - E[X_r X_s]$$

By hypothesis B, $\Delta_2 \longrightarrow 0$. It remains to show that $\Delta_1 \xrightarrow{P} 0$. By Chebychev's inequality,

$$P\{ |\Delta_1| \geq \varepsilon \} \leq \frac{E[\Delta_1^2]}{\varepsilon^2} = \frac{Var(\Delta_1)}{\varepsilon^2}$$

with

$$Var(\Delta_1) = \frac{Var(R_t)}{(n-q)} + \frac{2}{(n-q)^2} \sum_{k=1}^{n-q-1} (n-q-k) Cov(R_t, R_{t+k}) \qquad (1.9)$$

where by hypothesis A and B, Jensen's inequality and Fubini's theorem,

$$Var(R_t) = O\left( E\left( \int x_r x_s \delta_m(\vec{x}, \vec{X}_{t-1}) d\vec{x} \right)^2 \right) =$$

$$= O\left( E\left( \int x_r^2 x_s^2 \delta_m(\vec{x}, \vec{X}_{t-1}) d\vec{x} \right) \right) = O\left( \int x_r^2 x_s^2 E[\delta_m(\vec{x}, \vec{X}_{t-1})] d\vec{x} \right) =$$

$$= O\left( \int \left( \int x_r^2 x_s^2 \delta_m(\vec{x}, \vec{u}) d\vec{x} \right) f(\vec{u}) d\vec{u} \right) = O\left( \int u_r^2 u_s^2 f(\vec{u}) d\vec{u} \right) = O(1)$$

For the second terms in (1.9), we write

$$\gamma_R^m(1) = E[(R_t - R_{t,1})^2] = W_1 + W_2 + W_3$$

where $R_{t,1} = \int x_r x_s \delta_m(\vec{x}, \vec{X}_{t-1,1}) d\vec{x} - E[\int x_r x_s \delta_m(\vec{x}, \vec{X}_{t-1,1}) d\vec{x}]$ is the 1-$L^2$ approximation of $R_t$ ($\vec{X}_{t-1,1}$ is the 1-$L^2$ approximation of $\vec{X}_{t-1}$).

$$W_1 = E\left[\left( \int x_r x_s \delta_m(\vec{x}, \vec{X}_{t-1}) d\vec{x} - E\left[\int x_r x_s \delta_m(\vec{x}, \vec{X}_{t-1}) d\vec{x}\right] \right)^2\right]$$

$$W_2 = E\left[\left( \int x_r x_s \delta_m(\vec{x}, \vec{X}_{t-1,1}) d\vec{x} - E\left[\int x_r x_s \delta_m(\vec{x}, \vec{X}_{t-1,1}) d\vec{x}\right] \right)^2\right]$$

$$W_3 = -2E\left[\left(\int x_r x_s \delta_m(\vec{x},\vec{X}_{t-1})\,d\vec{x} - E\left[\int x_r x_s \delta_m(\vec{x},\vec{X}_{t-1})\,d\vec{x}\right]\right)\cdot\right.$$
$$\left.\cdot\left(\int x_r x_s \delta_m(\vec{x},\vec{X}_{t-1,l})\,d\vec{x} - E\left[\int x_r x_s \delta_m(\vec{x},\vec{X}_{t-1,l})\,d\vec{x}\right]\right)\right]$$

Application of hypotheses A and B to each $W_i$ shows that for fixed l and $n\to\infty$,

$$\gamma_R^m(l) = E\left[(X_r X_s)^2\right] + E\left[(X_{r,l} X_{s,l})^2\right] - 2E\left[(X_r X_s)(X_{r,l} X_{s,l})\right] + o(1)$$

and hence, by elementary calculations, that $\gamma_R^m(l) = O(\gamma(l)) + o(1)$, where $O(\cdot)$ refers to $l\to\infty$ and $o(\cdot)$ to $n\to\infty$.

Thus

$$|\mathrm{Cov}(R_t,R_{t+k})| \leq E\left[|R_t R_{t+k}|\right] = E\left[|(R_t - R_{t,l} + R_{t,l})R_{t+k}|\right] \leq$$
$$\leq E\left[|R_t - R_{t,l}|\,|R_{t+k}|\right] + E\left[|R_{t,l}|\,|R_{t+k} - R_{t+k,l} + R_{t+k,l}|\right] \leq$$
$$\leq (\gamma_R^m(l))^{\frac{1}{2}} O(1) + E\left[|R_{t,l} R_{t+k,l}|\right].$$

Now by the $L^2$-stability property of $\vec{Y}_t$, $\vec{X}_{t-1}$ is $L^2$-stable with respect to the q-dependent $\vec{\eta}_t$. (In particular $\vec{\eta}_t$ is $\phi$-mixing for $\phi$ such that for $k>q$ $\phi(k) = 0$). Thus both $\vec{X}_{t-1,l}$ and $R_{t,l}$ are $\phi$-mixing for $\phi_l$ such that for $k>l+q$ $\phi_l(k) = 0$, and hence, by Lemma 1 (Eq. 20.35) in Billingsley (1968),

$$E\left[|R_{t,l} R_{t+k,l}|\right] \leq 2(\phi_l(k))^{\frac{1}{2}} E\left[|R_{t,l}^2|\right].$$

Therefore,

$$P\{|\Delta_1| \geq \varepsilon\} \leq \frac{O(1)}{(n-q)} + O(1)\sum_{k=1}^{n-q-1} \frac{1}{(n-q)}\,(O(\gamma(l)) + o(1))^{\frac{1}{2}} +$$
$$+ O(1)\sum_{k=1}^{n-q-1} \frac{1}{(n-q)}\,(\phi_l(k))^{\frac{1}{2}} \qquad (1.10)$$

Balancing k and l in (1.10) with $l = [k/2]$ (i.e. the integer part of k/2) the third term is negligible, and Toeplitz' lemma shows that

$$\sum_{k=1}^{n-q-1} \frac{1}{(n-q)}\{\gamma([k/2])\}^{\frac{1}{2}} \longrightarrow 0$$

(since $\gamma(l)\to 0$ as $l\to\infty$ for AR(q)). Hence $\Delta_1 \xrightarrow{P} 0$. Similar reasoning proves (1.8).

<u>Proof of b)</u>

Imitating (1.6), (1.5) can be written in the form

$$\hat{\theta}_n = \left(\int \vec{x}\vec{x}^t\,d\hat{F}_q^n(\vec{x})\right)^{-1} \int \vec{x}y\,d\hat{F}_{q+1}^n(\vec{x},y)$$

where

$$\hat{F}^n_{q+1}(\vec{x},y) = \sum_{\{(\vec{X}_{i-1},X_i)/X_i \leq y\}} \frac{1}{(n-q)} \int_{-\infty}^{\vec{x}} \delta_m(\vec{X}_{i-1},\vec{t})\,d\vec{t} =$$

$$= \int_{(-\infty,-\infty)}^{(\infty,y)} \left( \int_{-\infty}^{\vec{x}} \delta_m(\vec{u},\vec{t})\,d\vec{t} \right) d\hat{F}^n_{q+1}(\vec{u},v)$$

and $\hat{F}^n_q(x) = \int_{-\infty}^{\vec{x}} \hat{f}_n(t)\,dt$, i.e. the $\Omega_n$ defined for (1.5).

By using Mallows metrics similar to those employed by Cristóbal et al (1986), it may be shown that

$$\hat{\theta}_n \xrightarrow{a.s.} \theta_0$$

if and only if

$$\int \|(\vec{x},y)\|^4 d\hat{F}^n_{q+1}(\vec{x},y) \xrightarrow{a.s.} \int \|(\vec{x},y)\|^4 dF_{q+1}(\vec{x},y) \quad \text{and}$$

$\hat{F}^n_{q+1} \longrightarrow F_{q+1}$ weakly almost sure.

To show the last we write $\hat{F}^n_{q+1}(\vec{x},y) - F_{q+1}(\vec{x},y) = \Delta_1 + \Delta_2 + \Delta_3$ where

$$\Delta_1 = \int_{(-\infty,-\infty)}^{(\infty,y)} \left( \int_{-\infty}^{\vec{x}} \delta_m(\vec{u},\vec{t})\,d\vec{t} \right) d(F^n_{q+1}(\vec{u},v) - F_{q+1}(\vec{u},v))$$

$$\Delta_2 = \int_{(\vec{x},-\infty)}^{(\infty,y)} \left( \int_{-\infty}^{\vec{x}} \delta_m(\vec{u},\vec{t})\,d\vec{t} \right) dF_{q+1}(\vec{u},v)$$

$$\Delta_3 = \int_{(-\infty,-\infty)}^{(\vec{x},y)} \left( \int_{-\infty}^{\vec{x}} \delta_m(\vec{u},\vec{t})\,d\vec{t} - 1 \right) dF_{q+1}(\vec{u},v)$$

Integrating by parts,

$$\Delta_1 = \int_{-\infty}^{+\infty} \left( \int_{-\infty}^{\vec{x}} \delta_m(\vec{u},\vec{t})\,d\vec{t} \right) d(F^n_{q+1}(\vec{u},y) - F_{q+1}(\vec{u},y))$$

and application of a Glivenko-Cantelli type theorem (Stute et al, 1980) shows that

$$\Delta_1 \xrightarrow{a.s.} 0.$$

Since $\Delta_2, \Delta_3 \to 0$ by the bounded convergence theorem, $\hat{F}^n_{q+1}$ almost surely converges weakly to $F_{q+1}$.

To show that

$\int \|(\vec{x},y)\|^4 d\hat{F}^n_{q+1}(\vec{x},y) \xrightarrow{a.s.} \int \|(\vec{x},y)\|^4 dF_{q+1}(\vec{x},y)$ we write

$$\Delta_1 = \frac{1}{(n-q)} \sum_{i=q+1}^{n} \int (\|\vec{x}\|^4 - \|\vec{X}_{i-1}\|^4)\, \delta_m(\vec{x}, \vec{X}_{i-1})\, d\vec{x}$$

$$\Delta_2 = \frac{2}{(n-q)} \sum_{i=q+1}^{n} X_i^2 \int (\|\vec{x}\|^2 - \|\vec{X}_{i-1}\|^2)\, \delta_m(\vec{x}, \vec{X}_{i-1})\, d\vec{x}$$

$$\Delta_3 = \frac{1}{(n-q)} \sum_{i=q+1}^{n} (\|\vec{X}_{i-1}\|^2 + X_i^2)^2 - E\big|\, \|(\vec{X}_{q-1}, X_q)\|^4 \big|$$

Since

$$\int \|(\vec{x},y)\|^4 d\hat{F}^n_{q+1}(\vec{x},y) - E\big|\,\|(\vec{X}_{q-1},X_q)\|^4\big| = \Delta_1 + \Delta_2 + \Delta_3$$

Using the condition c) $\Delta_1, \Delta_2 \xrightarrow{a.s} 0$. Finally, since

$\Delta_3 = \int \|(\vec{x},y)\|^4 d(F^n_{q+1}(\vec{x},y) - F_{q+1}(\vec{x},y))$, the boundedness of its support together results of Stute et al, (1980) implies that

$$\Delta_3 \xrightarrow{a.s.} 0,$$

and the proof of (b) is complete.

<u>Proof of c)</u>.

We write $\sqrt{n}(\hat{\theta}_n - \theta_0) = \sqrt{n}(\hat{\theta}_n - \tilde{\theta}_n) + \sqrt{n}(\tilde{\theta}_n - \theta_0)$, where $\tilde{\theta}_n$ is the least squares estimator defined in (1.3). The hypotheses made in the statement of (c) imply that

$$\sqrt{n}(\tilde{\theta}_n - \theta_0) \xrightarrow{d} N_q(0, \sigma^2 (E[\vec{X}_t \vec{X}_t^t])^{-1})$$

(see Fuller, 1976); and by the hypothesis made in the statement of (c) regarding $\delta_m$,

$$\sqrt{n}(\hat{\theta}_n - \tilde{\theta}_n) \xrightarrow{P} 0 ,$$

which concludes the proof. #

<u>Acknowledgements</u>. We wish to thank the Professor Wertz and the referees who made possible the best presentation of this paper.

## REFERENCES


BIERENS, H.J. (1983). Uniform consistency of kernel estimators of a regression function under generalized conditions. J. Amer. Statist. Assoc. 78, 699-707.

BILLINGSLEY, P. (1968). Convergence of Probability Measures. New York. John Wiley.

BOSQ, D. (1983). Non Parametric Prediction in Stationary Processes. Lecture Notes in Statistics 16, 69-84.

COLLOMB, G. (1982). Prédiction non paramétrique: étude de l'erreur quadratique du prédictogramme. C.R. Acad. Sc. Paris, t. 294, 59-62.

COLLOMB, G. and DOUKHAN, P. (1983). Estimation non paramétrique de la fonction d'autorégression d'un processus stationnaire et $\phi$-mélangeant: risques quadratiques pour le méthode du noyau. C.R. Acad. Sc. Paris, t. 296, 859-862.

COLLOMB, G. and HARDLE, W. (1986). Strong uniform convergence rates in robust nonparametric time series analysis and prediction: kernel regression estimation from dependent observations. Theory of Probability and its Applications. (To appear).

CRISTOBAL, J.A., FARALDO, P. and GONZALEZ MANTEIGA, W. (1986). A class of linear regression parameter estimators constructed by nonparametric estimation. (To appear in Ann. of Stat.).

FARALDO, P. and GONZALEZ MANTEIGA, W.(1986). On efficiency of a new class of linear regression estimates obtained by preliminary non-parametric estimation. New Perspectives in Theoretical and Applied Statistics. Puri. M. et al., Eds. John Wiley. (To appear).

FULLER, W. (1976). Introduction to statistical time series. New York. John Wiley.

HART. J.D. (1984). Efficiency of a kernel Density Estimator Under an Autoregressive Dependence Model. J. Amer. Statist. Assoc. 79, 111-117.

MASRY, E. (1983). Probability Density Estimation from Sampled Data. I.E.E.E. Transact. on Inform. Theory. 29. 696-709.

TITTERINGTTON, D.M.(1985). Common structure of smoothing techniques in Statistics. Inter. Stat. Review. 53, 141-170.

STUTE, W. and SCHUMANN, G. (1980). A general Glivenko-Cantelli Theorem for Stationary Sequences of Random Observations. Scand. J. Statist. 7, 102-104.

YAKOWITZ, S.J. (1985). Nonparametric Density Estimation, Prediction and Regression for Markov Sequences. J. Amer. Statist. Assoc. 80, 215-221.

W. González Manteiga
J.M. Vilar Fernández

Departamento de Estadística e Investigación Operativa
Facultad de Matemáticas

Universidad de Santiago de Compostela

E-15000 Santiago de Compostela
Spain

# SELECTING THE BEST UNKNOWN MEAN FROM NORMAL POPULATIONS HAVING A COMMON UNKNOWN COEFFICIENT OF VARIATION

Shanti S. Gupta and TaChen Liang
Department of Statistics
Purdue University
West Lafayette, Indiana 47907
USA

ABSTRACT. This paper deals with the problem of selecting the population associated with the largest unknown mean from several normal populations having a common unknown coefficient of variation. Both subset selection and indifference zone approaches are studied. Based on the observed sample means and sample standard deviations, a subset selection rule is proposed. Some properties related to this selection rule are discussed. For the indifference zone approach, a two-stage elimination type selection rule is considered. If the experimenter has some prior knowledge about an upper bound on the unknown means, a modification is introduced to reduce the size of the selected subset at the first stage and also to reduce the sample size at the second stage. An example is provided which indicates that the saving in the total sample size is quite significant if this prior knowledge is taken into consideration in designing the selection rule. It is shown how to implement the above selection rules by using several existing tables.

## 1. INTRODUCTION

The problem of selecting the population associated with the largest unknown mean of several normal populations (hereafter, referred to as the normal mean problem) has been extensively studied in the literature. Bechhofer (1954) proposed the so-called indifference zone approach to this problem and gave a single-stage selection rule for the case of known variance. For the case of unknown variances, two-stage selection rule have been studied by Bechhofer, Dunnett and Sobel (1954), Dudewicz and Dalal (1975), Rinott (1978), Mukhopadhyay (1979) and Gupta and Kim (1984), among others. Another so-called subset selection approach to the normal mean problem was proposed by Gupta (1956, 1965) who proposed a single-stage selection rule which is applicable to both cases, namely a common known variance and a common unknown variance. Also, Gupta and Wong (1982) studied a single-stage selection rule for the case of unknown and unequal variances using the subset selection approach. For detailed references on the normal mean problem, for both indifference zone and subset selection approaches, the reader is referred to Gupta



*P. Bauer et al. (eds.), Mathematical Statistics and Probability Theory, Vol. B, 97–112.*

and Panchapakesan (1979).

All of the work mentioned above on the normal mean problem assumed that the population means and the population variances are unrelated parameters. However, as pointed out by Amemiya (1973) and the references quoted there, it is quite common, particularly in biological and economic applications, that the population standard deviations are proportional to the population means. The constant of proportionality is referred to as the coefficient of variation. In such a situation, those procedures which have been developed for the case of unknown variance(s) might not be appropriate. The knowledge of the common coefficient of variation should be used to develop new selection rules. With this in mind, Tamhane (1978), by using the estimators developed by Gleser and Healy (1976), proposed selection rules (for both subset selection and indifference zone approaches) for normal populations having a common known coefficient of variation, and provided tables for implementing the rules in the large sample case. With the same assumption, Gupta and Singh (1983) proposed a subset selection rule, based on sample variances, for selecting the population associated with the largest mean. Their selection rule is independent of the value of the common coefficient of variation, and therefore, can be applied to the situation when the value of the common coefficient of variation is unknown.

In this paper, the problem of selecting the population with the largest mean from several normal populations having a common unknown coefficient of variation is studied. The statistical formulation of the problem is given in Section 2. In Section 3, a subset selection rule, based on the sample means and sample standard deviations, is proposed. Some properties related to this selection rule are also discussed. In Section 4, a two-stage elimination type rule is given based on the indifference zone approach. If the experimenter has some prior knowledge about an upper bound on the unknown means, a modification is introduced which reduces the size of the selected subset at the first stage and also reduces the sample size at the second stage. To implement the proposed selection rules, the related tables are available from Gupta (1963), and Gupta, Panchapakesan and Sohn (1985).

## 2. FORMULATION OF THE SELECTION PROBLEMS

Let $\pi_1,\ldots,\pi_k$ be k ($\geq$ 2) normal populations with unknown positive means $\theta_1,\ldots,\theta_k$ and variances $b^2\theta_1^2,\ldots,b^2\theta_k^2$, where b is a common unknown coefficient of variation. Let $\theta_{[1]} \leq \ldots \leq \theta_{[k]}$ be the ordered values of the means. We shall assume that the experimenter has no prior knowledge concerning the correct pairing between $\pi_i$ and $\theta_{[j]}$ $(1 \leq i, j \leq k)$. The population corresponding to $\theta_{[k]}$ will be referred to as the best population. Note that the parameter b is a nuisance parameter for our selection problem and it will be suppressed in the notation of the parameter space for simplicity. Therefore, in the following, we will let $\underset{\sim}{\theta} = (\theta_1,\ldots,\theta_k)$ be the parameter vector and $\Omega$ be the parameter space.

Subset Selection Approach

According to this approach, the goal of the experimenter is to select a (small) subset of populations which contains the best population. The selection of any subset containing the best population is called a correct selection (CS). The decision-maker restricts attention only to those rules which guarantee the probability requirement (the so-called P*-condition) that

$$P_{\underset{\sim}{\theta}}\{CS\} \geq P^* \quad \text{for all } \underset{\sim}{\theta} \in \Omega, \tag{2.1}$$

where $P^*$, $k^{-1} < P^* < 1$, is a preassigned constant.

Indifference Zone Approach

According to this approach, the goal of the experimenter is to select the best population. The selection of the best population is called the correct selection (CS). For this approach, in order to specify the probability requirement, it is first necessary to define a measure of distance between two populations. We consider the following measure of distance: $\delta(\theta_i,\theta_j) = \theta_i-\theta_j$, the difference between the two population means of $\pi_i$ and $\pi_j$. Note that this measure of distance is different from that considered by Tamhane (1978). For a preassigned value $\delta^* > 0$, let $\Omega(\delta^*) = \{\underset{\sim}{\theta} \in \Omega | \theta_{[k]} \geq \theta_{[k-1]} + \delta^*\}$. $\Omega(\delta^*)$ is known as the preference zone and its complement as indifference zone. The assignment of $\delta^*$ value will be based on the experimenter's prior knowledge. The experimenter restricts consideration only to those rules which guarantee the probability requirement that

$$P_{\underset{\sim}{\theta}}\{CS\} \geq P^* \quad \text{for all } \underset{\sim}{\theta} \in \Omega(\delta^*), \tag{2.2}$$

where $P^*$, $k^{-1} < P^* < 1$, is a preassigned constant.

## 3. A SUBSET SELECTION RULE

The goal here is to select, on the basis of an independent random sample $X_{ij}$, $j = 1,\ldots,n$, from each $\pi_i$ $(1 \leq i \leq k)$, a subset containing the best population with the minimum probability P* of a correct selection.

For each i, let $\bar{X}_i = \frac{1}{n}\sum_{j=1}^{n} X_{ij}$ and $S_i^2 = \frac{1}{n-1}\sum_{j=1}^{n}(X_{ij}-\bar{X}_i)^2$. Based on $(\bar{X}_i, S_i)$, $1 \le i \le k$, we propose a subset selection rule $R_1$ as follows:

$$R_1: \text{ Select } \pi_i \text{ if } \bar{X}_i \ge \bar{X}_j - \sqrt{\frac{2}{n}}\, cS_i \quad \text{for } j \ne i. \tag{3.1}$$

The constant c > 0 is determined by

$$P\{Z_j \le cW \quad \text{for } 1 \le j \le k-1\} = P^* \tag{3.2}$$

where

$$\begin{cases} Z_j \sim N(0,1), & 1 \le j \le k-1; \\ \operatorname{cov}(Z_i,Z_j) = 1/2, & \text{for } i \ne j; \\ (n-1)W^2 \sim \chi^2(n-1); & \text{and} \\ (Z_1,\ldots,Z_{k-1}) \text{ and } W \text{ are independent.} \end{cases} \tag{3.3}$$

The value of c satisfying (3.2) is available from Gupta and Sobel (1957), and Gupta, Panchapakesan and Sohn (1985).

Let $P_{\underset{\sim}{\theta}}\{CS|R\}$ be the probability of a correct selection when $\underset{\sim}{\theta}$ is the true parameter vector and the selection rule R is applied. Then, we have

<u>Theorem 3.1</u>. $P_{\underset{\sim}{\theta}}\{CS|R_1\} \ge P^*$ for all $\underset{\sim}{\theta} \in \Omega$.

<u>Proof</u>: Without loss of generality, we can assume that $\theta_k = \theta_{[k]}$.

Following a straightforward computation,

$$\begin{aligned} & P_{\underset{\sim}{\theta}}\{CS|R_1\} \\ &= P_{\underset{\sim}{\theta}}\{\bar{X}_k \ge \bar{X}_j - \sqrt{2/n}\, cS_k \quad \text{for } j \ne k\} \\ &= P_{\underset{\sim}{\theta}}\{Y_j \le \sqrt{n}b^{-1}\alpha_j(\underset{\sim}{\theta}) + \sqrt{2}c\beta_j(\underset{\sim}{\theta})W \quad \text{for } 1 \le j \le k-1\}, \end{aligned} \tag{3.4}$$

where

$$\begin{cases} Y_j = \sqrt{n}\,(\bar{X}_j-\bar{X}_k-\theta_j+\theta_k)b^{-1}(\theta_k^2+\theta_j^2)^{-\frac{1}{2}} \sim N(0,1), \quad 1 \le j \le k-1 \\ \operatorname{cov}(Y_i,Y_j) = \theta_k^2[(\theta_k^2+\theta_i^2)(\theta_k^2+\theta_j^2)]^{-\frac{1}{2}} \equiv \rho_{ij}(\underset{\sim}{\theta}), \quad i \ne j \\ W = S_k/(b\theta_k), \quad (n-1)W^2 \sim \chi^2(n-1) \quad \text{and} \\ (Y_1,\ldots,Y_{k-1}) \text{ and } W \text{ are independent;} \end{cases} \tag{3.5}$$

and

$$\begin{cases} \alpha_j(\underset{\sim}{\theta}) = (\theta_k-\theta_j)(\theta_k^2+\theta_j^2)^{-\frac{1}{2}}, \quad 1 \le j \le k-1; \\ \beta_j(\underset{\sim}{\theta}) = \theta_k(\theta_k^2+\theta_j^2)^{-\frac{1}{2}}, \quad 1 \le j \le k-1. \end{cases} \tag{3.6}$$

Note that $\alpha_j(\underset{\sim}{\theta}) \ge 0$, $\beta_j(\underset{\sim}{\theta}) \ge 1/\sqrt{2}$ and $\rho_{ij}(\underset{\sim}{\theta}) \ge \frac{1}{2}$ since $\theta_k \ge \theta_j$ for $j \ne k$. Then by Slepian's inequality, we conclude that the smallest value of the constant c is given by

$$P_{\underset{\sim}{\theta}}\{CS|R_1\} \ge P\{Z_j \le cW \text{ for } 1 \le j \le k-1\} = P^* \text{ for } \underset{\sim}{\theta} \in \Omega,$$

where $Z_j$ $(1 \le j \le k-1)$ are defined in (3.3).

Least Favorable Configuration

Let $\Omega_0 = \{\underset{\sim}{\theta}|\theta_1 = \ldots = \theta_k > 0\}$. For each $\underset{\sim}{\theta} \in \Omega$, $\theta_k = \theta_{[k]}$ being fixed, both $\alpha_j(\underset{\sim}{\theta})$ and $\beta_j(\underset{\sim}{\theta})$ are decreasing in $\theta_j$ and $\alpha_j(\underset{\sim}{\theta}) = 0$, $\beta_j(\underset{\sim}{\theta}) = 1/\sqrt{2}$ when $\theta_j = \theta_k$. Then by Slepian's inequality, we see that

$$\inf_{\underset{\sim}{\theta}\in\Omega} P_{\underset{\sim}{\theta}}\{CS|R_1\} = \inf_{\underset{\sim}{\theta}\in\Omega_0} P_{\underset{\sim}{\theta}}\{CS|R_1\} = P_{\underset{\sim}{\theta}_0}\{CS|R_1\}$$

for any $\underset{\sim}{\theta}_0 \in \Omega_0$. Note that $P_{\underset{\sim}{\theta}_0}\{CS|R_1\}$ does not depend on $\underset{\sim}{\theta}_0$.

Some Properties of $R_1$

<u>Property 1</u>. For fixed $\underset{\sim}{\theta} \in \Omega$, $P_{\underset{\sim}{\theta}}\{CS|R_1\}$ is decreasing in b for $b > 0$.

This is obvious from (3.4) since $\alpha_j(\underset{\sim}{\theta}) \geq 0$ and the distributions of $(Y_1,\ldots,Y_{k-1})$ and W are independent of the parameter b.

For fixed b, suppose that $(\theta_1,\ldots,\theta_k) = (ac_1 + d_1,\ldots,ac_k + d_k)$, where $c_i > 0$, a, $d_i \geq 0$ and $ac_i + d_i > 0$ for each $i = 1,\ldots,k$. Also, we assume that $c_k \geq c_j$, $d_k \geq d_j$ for $j \neq k$. Thus, $\theta_k \geq \theta_j$ for $j \neq k$. Then,

$$\begin{aligned} & P_{\underset{\sim}{\theta}}\{CS|R_1\} \\ & = P_{\underset{\sim}{\theta}}\{Y_j \leq \sqrt{n}\, g_j(a)b^{-1} + \sqrt{2}ch_j(a)W \quad \text{for } j \neq k\}, \end{aligned} \tag{3.7}$$

where $g_j(a) = [a(c_k-c_j)+(d_k-d_j)][(ac_k+d_k)^2 + (ac_j+d_j)^2]^{-\frac{1}{2}}$;

$$h_j(a) = (ac_k+d_k)[(ac_k+d_k)^2+(ac_j+d_j)^2]^{-\frac{1}{2}}.$$

<u>Lemma 3.1</u>. The following three statements are equivalent.

i) $\frac{d_k}{c_k} < \frac{d_j}{c_j}$.

ii) $g_j(a)$ is increasing in a for $a > 0$.

iii) $h_j(a)$ is increasing in a for $a > 0$.

From (3.7) and Lemma 3.1, we have the following results.

<u>Property 2</u>. Let b be a fixed constant, then

a) if $\frac{d_k}{c_k} > \frac{d_j}{c_j}$ for all $j \neq k$, then $P_{\underset{\sim}{\theta}}\{CS|R_1\}$ is decreasing in a; and

b) if $\frac{d_k}{c_k} < \frac{d_j}{c_j}$ for all $j \neq k$, then $P_{\underset{\sim}{\theta}}\{CS|R_1\}$ is increasing in a.

c) When $c_1 = c_2 = \ldots = c_k$ and $d_1,\ldots,d_k$ are fixed constants, then $P_{\underset{\sim}{\theta}}\{CS|R_1\}$ is decreasing in a. Note here that, under the assumption that $d_k \geq d_j$ for all $j \neq k$, the population associated with $d_k$ is the best population.

d) When $d_1 = \ldots = d_k = 0$ and $c_1, \ldots, c_k$ are fixed constants, then $P_{\underset{\sim}{\theta}}\{CS|R_1\}$ is a constant, which is independent of the parameter a.

<u>Property 3. Monotonicity of $R_1$</u>. Let $P_{\underset{\sim}{\theta}}\{\pi_i|R_1\}$ be the probability of including $\pi_i$ in the selected subset when rule $R_1$ is applied and $\underset{\sim}{\theta}$ is the true parameter vector. Then,

<u>Theorem 3.2</u>. If $\theta_i \geq \theta_j$, then $P_{\underset{\sim}{\theta}}\{\pi_i|R_1\} \geq P_{\underset{\sim}{\theta}}\{\pi_j|R_1\}$.

<u>Proof</u>: Without loss of generality, let $\theta_2 \geq \theta_1$, and we will prove that $P_{\underset{\sim}{\theta}}\{\pi_2|R_1\} \geq P_{\underset{\sim}{\theta}}\{\pi_1|R_1\}$. Straightforward computations show that

$$P_{\underset{\sim}{\theta}}\{\pi_i|R_1\}$$

$$= P_{\underset{\sim}{\theta}}\left\{\begin{array}{l} Y_{ij} \leq \sqrt{n}(\theta_i-\theta_j)b^{-1}(\theta_i^2+\theta_j^2)^{-\frac{1}{2}}+\sqrt{2}c\theta_i(\theta_i^2+\theta_j^2)^{-\frac{1}{2}}W \text{ and} \\ Y_{im} \leq \sqrt{n}(\theta_i-\theta_m)b^{-1}(\theta_i^2+\theta_m^2)^{-\frac{1}{2}}+\sqrt{2}c\theta_i(\theta_i^2+\theta_m^2)^{-\frac{1}{2}}W \text{ for} \\ \hfill 3 \leq m \leq k \end{array}\right\}$$

for $i,j = 1,2$, $i \neq j$, where $\text{cov}(Y_{im}, Y_{i\ell}) = \theta_i^2[(\theta_i^2+\theta_m^2)(\theta_i^2+\theta_\ell^2)]^{-\frac{1}{2}}$, $1 \leq \ell, m \leq k$, $\ell \neq m$, $\ell \neq i$, $m \neq i$.

We then conclude the proof of this theorem by an application of Slepian's inequality and by noting the following facts:

i) $(\theta-\theta_m)(\theta^2+\theta_m^2)^{-\frac{1}{2}}$ is increasing in $\theta$ for $\theta > 0$ for each m;

ii) $\theta(\theta^2+\theta_m^2)^{-\frac{1}{2}}$ is increasing in $\theta$ for $\theta > 0$ for each m; and

iii) $\text{cov}(Y_{2m}, Y_{2\ell}) \geq \text{cov}(Y_{1m}, Y_{1\ell})$ for $3 \leq m, \ell \leq k$, $m \neq \ell$ and

$\text{cov}(Y_{21}, Y_{2m}) \geq \text{cov}(Y_{12}, Y_{1m})$ for $3 \leq m \leq k$.

Expected Size of the Selected Subset

Note the selection rule $R_1$ selects a non-empty subset, the size of which is not fixed in advance but depends on the outcome of the experiment. Hence, as a measure of the performance of the selection rule $R_1$, we can consider the expected size of the selected subset, say $E_{\underset{\sim}{\theta}}(S|R_1)$. We have the following expression:

$$\begin{aligned} E_{\underset{\sim}{\theta}}(S|R_1) &= \sum_{i=1}^{k} P_{\underset{\sim}{\theta}}\{\pi_i|R_1\} \\ &= \sum_{i=1}^{k} P_{\underset{\sim}{\theta}}\{V_i \geq V_j\theta_j\theta_i^{-1}+\sqrt{n}b^{-1}(\theta_j\theta_i^{-1}-1)-\sqrt{2}cW,\ j \neq i\}, \end{aligned} \tag{3.8}$$

where $V_i$, $1 \leq i \leq k$, are iid having standard normal distribution and $W$ is as defined in (3.3). It is often of interest to identify the parameter configuration where the supremum of (3.8) occurs. For $k = 2$, let $\theta_2 = \theta$ and $\theta_1 = \Delta\theta$, $0 < \Delta \leq 1$. Then,

$$\begin{aligned} E_{\underset{\sim}{\theta}}(S|R_1) &= \int_0^\infty \Phi(\sqrt{n}b^{-1}(\Delta-1)(1+\Delta^2)^{-\frac{1}{2}}+\sqrt{2}\Delta c(1+\Delta^2)^{-\frac{1}{2}}w)\,dF_W(w) \\ &\quad + \int_0^\infty \Phi(\sqrt{n}b^{-1}(1-\Delta)(1+\Delta^2)^{-\frac{1}{2}}+\sqrt{2}c(1+\Delta^2)^{-\frac{1}{2}}w)\,dF_W(w), \end{aligned}$$

where $\Phi(\cdot)$ is the standard normal distribution and $F_W(\cdot)$ is the distribution of the random variable $W$.

We see that $\sup_{\underset{\sim}{\theta}\in\Omega} E_{\underset{\sim}{\theta}}(S|R_1) = 2P^*$ and the supremum occurs when $\Delta = 1$. For $k > 2$, it appears difficult to obtain such a result. We can only state that
$\sup_{\underset{\sim}{\theta}\in\Omega} E_{\underset{\sim}{\theta}}(S|R_1) \geq \sup_{\underset{\sim}{\theta}\in\Omega_0} E_{\underset{\sim}{\theta}}(S|R_1) = kP^*$.

Remark

For the case when b is known, Tamhane (1978), proposed a subset selection rule for the best population using estimators of Gleser and Healy (1976). He also provided tables for implementing the rule in the large sample case. However, for small sample sizes, tables for implementing Tamhane's rule are not available. It has been pointed out by Tamhane (1978) that for certain values of b, k, n and P*, Tamhane's rule does not exist. Based on sample variances, Gupta and Singh (1983) also proposed a subset selection rule for the problem of selecting the best population. Their selection rule and the associated probability of the correct selection are independent of the value of the common coefficient of variation b, and hence, can be applied to the situation when this value is unknown. They made some comparison between Tamhane's and their rules. It is found that in terms of expected size of selected subset, the performance of Gupta and Singh's rule is a little inferior to that of Tamhane's rule. For k = 2 and large sample size, there is not much difference between these two rules.

Note that the proposed rule $R_1$ is independent of the value of the common coefficient of variation b. However, from (3.4), we can see that the associated probability of a correct selection depends on b for each fixed $\underset{\sim}{\theta} \in \Omega$. Therefore, it is interesting to compare the performance of the rule $R_1$ with Gupta and Singh's rule. Further study of this comparison is to be carried out.

## 4. INDIFFERENCE ZONE APPROACH

The goal here is to derive selection rules which will select the best population with a guaranteed probability P*. On $\Omega(\delta^*)$, the associated measure of distance between populations $\pi_i$ and $\pi_j$ is $\delta(\theta_i,\theta_j) = \theta_i-\theta_j$, which is different from the one considered by Tamhane (1978). Since both $\underset{\sim}{\theta}$ and the common coefficient of variation b are unknown, on $\Omega(\delta^*)$, it is impossible to construct a single-stage selection rule which guarantees the probability requirement of (2.2). In the following, a two-stage elimination type selection rule is proposed.

Two-Stage Elimination Type Selection Rule $R_2$

<u>Stage 1</u>: Take $n_0(\geq 2)$ independent observations $X_{ij}(j = 1,\ldots,n_0)$ from each $\pi_i$ $(i = 1,\ldots,k)$, and compute the sample mean $\bar{X}_i^{(1)} = \frac{1}{n_0}\sum_{j=1}^{n_0} X_{ij}$ and sample variance $S_i^2 = \frac{1}{n_0-1}\sum_{j=1}^{n_0}(X_{ij}-\bar{X}_i^{(1)})^2$. Then determine the set

$$A = \{i|\bar{X}_i^{(1)} \geq \bar{X}_j^{(1)} - (\sqrt{2}dS_i/\sqrt{n_0}-\delta^*)^+ \text{ for } j \neq i\}, \tag{4.1}$$

where $y^+ = \max(y,0)$ and d is a positive constant chosen to satisfy (2.2).

If A contains only one element, then stop sampling and assert that the population associated with $\max_{1\le j\le k} \bar{X}_j^{(1)}$ is the best.

If A has more than one element, then proceed to the second stage.

Stage 2: Let $S_A = \max_{i\in A} S_i$. Determine $N = \max\{n_0, [2(dS_A/\delta^*)^2]^*\}$, where $[y]^*$ denotes the smallest integer $\geq y$. Take $N-n_0$ additional observations $X_{ij}$ from each $\pi_i$ $(i \in A)$ if necessary. Then compute the overall sample means $\bar{X}_i = \frac{1}{N}\sum_{j=1}^{N} X_{ij}$ $(i \in A)$ and assert that the population associated with $\max_{i\in A} \bar{X}_i$ is the best.

Note: This selection rule is essentially of the same type as that of Gupta and Kim (1984), even though we have a different type of screening procedure and different way to determine the value N. The difference is due to the fact that in this paper, the concerned population standard deviations are proportional to respective population means while Gupta and Kim (1984) considered a common unknown standard deviation.

Probability of Correct Selection

Without loss of generality, we still assume that $\pi_k$ is the best population. Therefore under the preference zone $\theta_k \geq \theta_j + \delta^*$ for $j \neq k$.

Let B be any subset of $\{1,2,\ldots,k\}$ containing the element k; and let E(B) be the event that subset B is selected at the first stage. Also, let $\mathcal{C} = \{B \subset \{1,\ldots,k\} \mid k \in B\}$. Then,

$$\bigcup_{B\in\mathcal{C}} E(B) = \{\bar{X}_k^{(1)} \geq \bar{X}_j^{(1)} - (\frac{\sqrt{2}dS_k}{\sqrt{n_0}} - \delta^*)^+ \text{ for } j \neq k\}. \tag{4.2}$$

Also,

$$\begin{aligned} P_{\underset{\sim}{\theta}}\{CS \mid R_2\} &= \sum_{B\in\mathcal{C}} P_{\underset{\sim}{\theta}}\{CS \cap E(B)\} \\ &= \sum_{B\in\mathcal{C}} P_{\underset{\sim}{\theta}}\{CS \mid E(B)\} P_{\underset{\sim}{\theta}}\{E(B)\}. \end{aligned} \tag{4.3}$$

Here $P_{\underset{\sim}{\theta}}\{CS|E(B)\}$ denotes the conditional probability of CS given E(B).

Since $\theta_k \geq \theta_j + \delta^*$ for $j \neq k$, for each $B \in \mathcal{C}$,

$$\begin{aligned} P_{\underset{\sim}{\theta}}\{CS|E(B)\} &= P_{\underset{\sim}{\theta}}\{\bar{X}_k \geq \bar{X}_j \text{ for } j \in B-\{k\}|E(B)\} \\ &\geq P_{\underset{\sim}{\theta}}\{Y_j \leq \sqrt{N}\delta^* b^{-1}(\theta_j^2+\theta_k^2)^{-\frac{1}{2}} \text{ for } j \in B-\{k\}|E(B)\}, \end{aligned} \tag{4.4}$$

where

$$Y_j = \sqrt{N}[(\bar{X}_j-\bar{X}_k)-(\theta_j-\theta_k)](b(\theta_j^2+\theta_k^2)^{\frac{1}{2}})^{-1} \sim N(0,1),\ 1 \leq j \leq k-1;$$

$\operatorname{cov}(Y_i,Y_j) = \rho_{ij}(\underset{\sim}{\theta})$ which is defined in (3.5) for $i \neq j$.

Conditional on E(B), $\sqrt{N}\delta^* \geq \sqrt{2}dS_B \geq \sqrt{2}dS_k$. Then, it follows that

$$\begin{aligned} P_{\underset{\sim}{\theta}}\{CS|E(B)\} &\geq P_{\underset{\sim}{\theta}}\{Y_j \leq \sqrt{2}\beta_j(\underset{\sim}{\theta})dW \text{ for } j \in B-\{k\}|E(B)\} \\ &\geq P_{\underset{\sim}{\theta}}\{Y_j \leq \sqrt{2}\beta_j(\underset{\sim}{\theta})dW \text{ for } 1 \leq j \leq k-1|E(B)\} \end{aligned} \tag{4.5}$$

where $W = S_k/(b\theta_k)$ and $\beta_j(\underset{\sim}{\theta})$ is defined in (3.6).

Let $Y_j^{(1)} = \sqrt{n_0}[(\bar{X}_j^{(1)}-\bar{X}_k^{(1)})-(\theta_j-\theta_k)](b(\theta_j^2+\theta_k^2)^{\frac{1}{2}})^{-1}$, $1 \leq j \leq k-1$.

Then, $Y_j^{(1)} \sim N(0,1)$, $\operatorname{cov}(Y_i^{(1)},Y_j^{(1)}) = \rho_{ij}(\underset{\sim}{\theta})$, $i \neq j$. Also,

$\operatorname{cov}(Y_i,Y_j^{(1)}) \geq 0$, $1 \leq i, j \leq k-1$. It should be pointed out that

$(Y_1,\ldots,Y_{k-1})$ and $(Y_1^{(1)},\ldots,Y_{k-1}^{(1)})$ are identically distributed, and

$(Y_1,\ldots,Y_{k-1})$ and $(Y_1^{(1)},\ldots,Y_{k-1}^{(1)})$ are independent of W.

For fixed $\theta_k (= \theta_{[k]})$, let $\underset{\sim}{\theta}^k = (\theta_1^k,\ldots,\theta_k^k)$, where $\theta_k^k = \theta_k$ and

$\theta_j^k = \theta_k - \delta^*$ for $j \neq k$. Then,

$$\begin{cases} \rho_{ij}(\underset{\sim}{\theta}) \geq \rho_{ij}(\underset{\sim}{\theta}^k); \\ \beta_j(\underset{\sim}{\theta}) \geq \beta_j(\underset{\sim}{\theta}^k). \end{cases} \tag{4.6}$$

Note that both $\rho_{ij}(\underset{\sim}{\theta}^k)$ and $\beta_j(\underset{\sim}{\theta}^k)$ are decreasing in $\theta_k$;

$\rho_{ij}(\underset{\sim}{\theta}^k) \to \frac{1}{2}$, $\beta_j(\underset{\sim}{\theta}^k) \to \frac{1}{\sqrt{2}}$ as $\theta_k \to \infty$.

From Equations (4.2) - (4.6) together with the facts that $\theta_k \geq \theta_j + \delta^*$ for $j \neq k$, $\mathrm{cov}(Y_i, Y_j^{(1)}) \geq 0$, $\sqrt{n_0}\,\delta^* + (\sqrt{2}dS_k - \sqrt{n_0}\delta^*)^+ \geq \sqrt{2}dS_k$ and repeated applications of Slepian's inequality lead to the following result:

$$\begin{aligned}
&P_{\underset{\sim}{\theta}}\{CS|R_2\} \\
&\geq P_{\underset{\sim}{\theta}}\{Y_j \leq \sqrt{2}\beta_j(\underset{\sim}{\theta})dW,\ Y_j^{(1)} \leq \sqrt{2}\beta_j(\underset{\sim}{\theta})dW \text{ for } 1 \leq j \leq k-1\} \\
&= \int_0^\infty P_{\underset{\sim}{\theta}}\{Y_j \leq \sqrt{2}\beta_j(\underset{\sim}{\theta})dw,\ Y_j^{(1)} \leq \sqrt{2}\beta_j(\underset{\sim}{\theta})dw \text{ for } 1 \leq j \leq k-1\}dF_W(w) \\
&\geq \int_0^\infty [P_{\underset{\sim}{\theta}}\{Y_j \leq \sqrt{2}\beta_j(\underset{\sim}{\theta})dw \text{ for } 1 \leq j \leq k-1\}]^2 dF_W(w) \\
&\geq \int_0^\infty [P_{\underset{\sim}{\theta}^k}\{Y_j \leq \sqrt{2}\beta_j(\underset{\sim}{\theta}^k)dw \text{ for } 1 \leq j \leq k-1\}]^2 dF_W(w) \\
&\geq [\int_0^\infty P_{\underset{\sim}{\theta}^k}\{Y_j \leq \sqrt{2}\beta_j(\underset{\sim}{\theta}^k)dw \text{ for } 1 \leq j \leq k-1\}dF_W(w)]^2 \\
&\geq [\int_0^\infty P\{Z_j \leq dw \text{ for } 1 \leq j \leq k-1\}dF_W(w)]^2 \\
&= [P\{Z_j \leq dW \text{ for } 1 \leq j \leq k-1\}]^2
\end{aligned} \tag{4.7}$$

where $(Z_1,\ldots,Z_{k-1})$ are defined in (3.3), and distributed independently of W.

For a given P*, we can choose the value d so that

$$P\{Z_j \leq dW \text{ for } 1 \leq j \leq k-1\} = \sqrt{P^*}. \tag{4.8}$$

Therefore, the probability requirement of (2.2) will be satisfied. For some specified values of P*, k and $n_0$, the corresponding d-values can be found in Gupta and Sobel (1957) and Gupta, Panchapakesan and Sohn (1985).

A Modified Two-Stage Selection Rule $R_2(\theta^*)$

In practical situations, the experimenter sometimes may have some prior knowledge about an upper bound on $\theta_{[k]}$, say $\theta^*$. This knowledge can be used to reduce the sample size taken at the second stage.

Let $\underset{\sim}{\theta}^*(\delta^*) = (\theta_1^*,\dots,\theta_k^*)$ where $\theta_{[k]}^* = \theta^*$ and $\theta_{[j]}^* = \theta^*-\delta^*$ for $j \neq k$. Let $\rho^* = \theta^{*2}(\theta^{*2} + (\theta^*-\delta^*)^2)^{-1}$ and $\beta^* = \theta^*(\theta^{*2} + (\theta^*-\delta^*)^2)^{-\frac{1}{2}}$. Let $Y_i^*(1 \leq i \leq k-1)$ be standard normal random variables having $\operatorname{cov}(Y_i^*,Y_j^*) = \rho^*$ for $i \neq j$, and W be a random variable distributed independently of $(Y_1^*,\dots,Y_{k-1}^*)$ with $(n_0-1)W^2$ having a $\chi^2$ distribution with $(n_0-1)$ degrees of freedom.

Let $d(\theta^*)$ be chosen to satisfy

$$P\{Y_i^* \leq d(\theta^*)\sqrt{2}\beta^* W \text{ for } 1 \leq j \leq k-1\} = \sqrt{P^*}. \tag{4.9}$$

Then a modified two-stage elimination type selection rule, say $R_2(\theta^*)$, can be defined. This selection rule $R_2(\theta^*)$ is similar to rule $R_2$. The only difference is that now the value $d(\theta^*)$ is used instead of the value d. We denote the corresponding N by $N(\theta^*)$.

Following (4.7), one can see that for each $\underset{\sim}{\theta} \in \Omega(\delta^*,\theta^*)$ where $\Omega(\delta^*,\theta^*) = \{\underset{\sim}{\theta} \in \Omega(\delta^*)\,|\,\theta_{[k]} \leq \theta^*\}$,

$$P_{\underset{\sim}{\theta}}\{CS|R_2(\theta^*)\} \geq [P\{Y_i^* \leq d(\theta^*)\sqrt{2}\beta^* W \text{ for } 1 \leq j \leq k-1\}]^2 = P^*.$$

Since $\rho^* > \frac{1}{2}$ and $\beta^* > \frac{1}{\sqrt{2}}$, it follows from Slepian's inequality, (4.8) and (4.9) that $d(\theta^*) < d$, and hence $N(\theta^*) \leq N$. Also, if we let $A(\theta^*)$ denote the subset selected at the first stage by applying the rule $R_2(\theta^*)$, that is,

$$A(\theta^*) = \{i|\bar{X}_i^{(1)} \geq \bar{X}_j^{(1)} - (\frac{\sqrt{2}d(\theta^*)S_i}{\sqrt{n_0}} - \delta^*)^+ \text{ for } j \neq i\}, \tag{4.10}$$

then, since $d(\theta^*) < d$, from (4.1) and (4.10), we can see that $A(\theta^*) \subset A$; hence, $|A(\theta^*)| \leq |A|$ where $|B|$ denote the size of set B.

Generally speaking, the modified selection rule $R_2(\theta^*)$ reduces the size of the subset selected at the first stage and it also reduces the sample size needed at the second stage.

An Example to Illustrate the Use of $R_2$ and $R_2(\theta^*)$

Suppose that a consumer has to decide on buying one lot of bolts from among five lots that are available. The tensile strength (in pound per square inch; psi) of the ith ($1 \leq i \leq 5$) lot is normally distributed with mean $\theta_i$ and standard deviation $b\theta_i$, where both b and $\theta_i$ are positive and unknown. Suppose that $\delta^* = 200$ psi and $P^* = 0.90$ have been specified by the consumer. Further, suppose that $n_0 = 16$ bolts are randomly sampled from each of the five lots and that the sample means and the sample standard deviations are:

$$(\bar{x}_1^{(1)}, \bar{x}_2^{(1)}, \bar{x}_3^{(1)}, \bar{x}_4^{(1)}, \bar{x}_5^{(1)}) = (350,380,470,600,650),$$

$$(s_1, s_2, s_3, s_4, s_5) = (360,420,500,580,600).$$

Now $\sqrt{P^*} = 0.9486833 \approx 0.95$, so from Table IV ($\rho = 0.5$) in Gupta, Panchapakesan and Sohn (1985), using interpolation, it is found that $d \approx 2.34582$. Then, $A = \{3, 4, 5\}$. Therefore, we proceed to the second stage and find that $N = 100$. Further additional 84 observations are taken from each of the selected lot and the sample mean $\bar{X}_i$ is computed. Finally, the consumer selects the lot of bolts associated with the largest sample mean among $\bar{X}_i$, $3 \leq i \leq 5$, as the best.

Suppose now that the consumer, from past experience, has some knowledge that $\theta_{[5]} \leq 1000 = \theta^*$. Therefore, he prefers to apply the modified selection rule $R_2(\theta^*)$ for his selection problem. Now,

$$\rho^* = \theta^{*2}(\theta^{*2} + (\theta^*-\delta^*)^2)^{-1} = 0.609756;$$

$$\beta^* = \theta^*(\theta^{*2} + (\theta^*-\delta^*)^2)^{-\frac{1}{2}} = 0.7808688.$$

Let $d_1 = d(\theta^*)\sqrt{2}\beta^*$. From (4.9) and Table IV ($\rho = 0.6$) in Gupta, Panchapakesan and Sohn (1985), using interpolation, it is found that

$d_1 = 2.30289$. Therefore $d(\theta^*) \approx 2.0853556$. Note that $\rho^* \approx 0.609756 > 0.6$. So the value $d(\theta^*)$ obtained in this way will be a little conservative since the exact value of $d(\theta^*)$ will be a little less than that used here. We then find that $A(\theta^*) = \{4,5\}$ and $N(\theta^*) = 79$. Therefore the consumer needs to take additional 63 observations from each of 4th and 5th lots to accomplish the selection process. Note that the total sample size by applying rule $R_2(\theta^*)$ is 206 while the total sample size by applying the rule $R_2$ is 332. The saving of the total sample size is quite significant.

## 5. ACKNOWLEDGEMENTS

This research was partially supported by the Office of Naval Research Contract N00014-84-C-0167 and NSF Grant DMS-8606964 at Purdue University.

## 6. REFERENCES

Amemiya, T. (1973). 'Regression analysis when the variance of the dependent variable is proportional to the square of its expectation'. J. Amer. Statist. Assoc. 68, 928-934.

Bechhofer, R. E. (1954). 'A single sample multiple decision procedure for ranking means of normal populations with known variances', Ann. Math. Statist. 25, 16-39.

Bechhofer, R. E., Dunnett, C. W. and Sobel, M. (1954). 'A two-sample multiple decision procedure for ranking means of normal populations with a common unknown variance', Biometrika 41, 170-176.

Dudewicz, E. J. and Dalal, S. R. (1975). 'Allocation of observations in ranking and selection with unequal variances', Sankhyā B37, 28-78.

Gleser, L. J. and Healy, J. D. (1976). 'Estimating the mean of a normal distribution with known coefficient of variation', J. Amer. Statist. Assoc. 71, 977-981.

Gupta, S. S. (1956). 'On a decision rule for a problem in ranking means', Institute of Statistics Mimeo. Ser. No. 150, University of North Carolina, Chapel Hill, N.C.

Gupta, S. S. (1963). 'Probability integrals of multivariate normal and multivariate t', Ann. Math. Statist. 34, 792-828.

Gupta, S. S. (1965). 'On some multiple decision (selection and ranking) rules', Technometrics 7, 225-245.

Gupta, S. S. and Kim, W. C. (1984). 'A two-stage elimination type procedure for selecting the largest of several normal means with a common unknown variance', Design of Experiments: Ranking and Selection, (Eds. T. J. Santner and A. C. Tamhane), Marcel Dekker, New York, 77-94.

Gupta, S. S. and Panchapakesan, S. (1979). Multiple Decision Procedures: Methodology of Selection and Ranking Populations, John Wiley, New York.

Gupta, S. S., Panchapakesan, S. and Sohn, J. K. (1985). 'On the distribution of the studentized maximum of equally correlated normal random variables', Commun. Statist.-Simula. Computa. 14(1), 103-135.

Gupta, S. S. and Singh, A. K. (1983). 'On subset selection procedures for the largest mean from normal populations having a common known coefficient of variation', Technical Report #83-6, Department of Statistics, Purdue University, West Lafayette, Indiana.

Gupta, S. S. and Sobel, M. (1957). 'On a statistic which arises in selection and ranking problem', Ann. Math. Statist. 28, 957-967.

Gupta, S. S. and Wong, W. Y. (1982). 'Subset selection procedures for the means of normal populations with unequal variances: unequal sample sizes case', Selecta Statistica Canadiana Vol. VI, 109-149.

Mukhopadhyay, N. (1979). 'Some comments on two-stage selection procedures', Commun. Statist. A8, 671-683.

Rinott, Y. (1978). 'On two-stage selection procedures and related probability inequalities', Commun. Statist. A7, 799-811.

Tamhane, A. C. (1978). 'Ranking and selection problems for normal populations with common known coefficient of variation', Sankhyā B39, 344-361.

# THE EXTREME LINEAR PREDICTIONS OF THE MATRIX-VALUED STATIONARY STOCHASTIC PROCESSES

B. Gyires
Kossuth L. University, Debrecen
P.O. Box 12
4010 Debrecen
Hungary

SUMMARY. Let $A$ be an arbitrary matrix with complex entries. It is known that the arithmetic mean of the diagonal elements of $AA^*$ is used for the measure of error of the linear predictions of matrix-valued stationary stochastic processes. It can be raised the question what happens if we apply another means of these diagonal elements. In this paper we use the geometric and harmonic means besides the arithmetic one for that purpose.

## I. PRELIMINARY

Denote by $C_p$ the set of the $p \times p$ matrices with complex entries. Let $C_p(H) \subset C_p$ be the set of the Hermite symmetric positive semidefinite matrices. By $C_p^+(H) \subset C_p(H)$ is denoted the set of the Hermite symmetric positive definite matrices.

Let $A = U \Lambda U^*$ be the spectral representation of $A \in C_p^+(H)$, where $\Lambda$ is a diagonal matrix with positive diagonal elements $\lambda_j$ $(j=1,\dots,p)$, and $UU^* = E$ with unit matrix $E$. Let $X \in C_p$ be an arbitrary regular matrix and let the diagonal elements of $(XU)^*(XU)$ be $s_j$ $(j=1,\dots,p)$.

We use the following notations

$$A(a_j\ (j=1,\dots,p)),\quad G(a_j\ (j=1,\dots,p)),$$
$$H(a_j\ (j=1,\dots,p))$$



P. Bauer et al. (eds.), Mathematical Statistics and Probability Theory, Vol. B, 113–124.

for the arithmetic, geometric and harmonic means of the numbers $a_j > 0$ $(j=1,\dots,p)$, respectively.

We introduce the following definitions:

$$A(XAX^*) = A(s_j\lambda_j \ (j=1,\dots,p)) , \tag{1.1}$$

$$G(XAX^*) = G(s_j\lambda_j \ (j=1,\dots,p)) , \tag{1.2}$$

$$H(XAX^*) = H(s_j\lambda_j \ (j=1,\dots,p)) . \tag{1.3}$$

It is easy to see that $A(XAX^*) = A(X^*XA)$. A similar property doesn't hold for the functionals $G$ and $H$.

Let

$$D = \{X \in C_p \mid \mathrm{Det}(X^*X) = 1\} ,$$

$$P = \{X \in C_p \mid \mathrm{Per}(X^*X) = 1\} ,$$

where $\mathrm{Det}(F)$, and $\mathrm{Per}(F)$ denote the determinant, and the permanent of $F \in C_p$, respectively.

<u>Theorem 1.1.</u> Let $A = U\Lambda U^*$ be the spectral representation of $A \in C_p^+(H)$. If $X \in D$ then

$$A(XAX^*) \geqq G(XAX^*) \geqq (\mathrm{Det}\ A)^{1/p} , \tag{1.4}$$

moreover if $XU \in P$ then

$$H(XAX^*) \leqq G(XAX^*) \leqq (\mathrm{Det}\ A)^{1/p} . \tag{1.5}$$

Equality is in all places if and only if

$$X^*X = (\mathrm{Det}\ A)^{1/p} A^{-1} . \tag{1.6}$$

<u>Proof.</u> The first inequality of (1.4), and (1.5), respectively, hold evidently by the well-known inequalities among The arithmetic, geometric and harmonic means.
there are equalities at both places if and only if

$$s_1\lambda_1 = \ldots = s_p \lambda_p , \tag{1.7}$$

where $s_j$ $(j=1,\ldots,p)$ are the diagonal elements of $(XU)^*(XU)$ , and $\lambda_j$ $(j=1,\ldots,p)$ denote the eigenvalues of $A$ .

In order to prove the second inequality of (1.4), and of (1.5), respectively, using the determinantal theorem of Hadamard ([7]), and the permanental theorem of Marcus ([5], [6]) we get that

$$s_1 \ldots s_p \geqq \mathrm{Det}(X^*X) = 1 ,$$

and that

$$s_1 \ldots s_p \leq \mathrm{Per}[(XU)^*(XU)] = 1$$

respectively, with equality in both expression if and only if $(XU)^*(XU)$ is a digonal matrix, i.e.

$$s_1 \ldots s_p = 1 . \tag{1.8}$$

By (1.7) and (1.8) we have

$$(\lambda_j s_j)^p = \lambda_1 \ldots \lambda_p ,$$

i.e.

$$s_j = \frac{1}{\lambda_j} (\mathrm{Det}\ A)^{1/p} \quad (j = 1,\ldots,p) ,$$

which is equivalent to (1.6).

By Theorem 1.1. the following statement holds.

<u>Theorem 1.2.</u> Let $A = U\Lambda U^*$ be the spectral representation of $A \in C_p^+(H)$ . Then

$$\inf_{X\in D} A(XAX^*) = \inf_{X\in D} G(XAX^*) = \sup_{XU\in P} G(XAX^*) =$$

$$= \sup_{XU\in P} H(XAX^*) = (\mathrm{Det}\ A)^{1/p} .$$

We shall use the following more general form of the Lagrange transformation ([1], Lemma 1).

Theorem 1.3. Let the block matrix

$$A = (a_{jk})_o^n$$

with blocks

$$a_{jk} \in C_p \qquad (j,k = 0,1,\dots,n)$$

be Hermite symmetric positive definite. Then the representation

$$A = B \begin{pmatrix} \gamma_o & & & \\ & \gamma_1 & & (0) \\ (0) & & \ddots & \\ & & & \gamma_n \end{pmatrix} B^* \qquad (1.9)$$

holds, where the matrices

$$\gamma_j \in C_p^+(H) \quad (j=0,1,\dots,n),$$

and the entries

$$\beta_{jk} \in C_p \qquad (j=0,1,\dots,k-1;\ k=1,\dots,n)$$

of the block matrix

$$B = \begin{pmatrix} E & O & O & \dots & O \\ \beta_{01} & E & O & \dots & O \\ \beta_{02} & \beta_{12} & E & \dots & O \\ \cdot & \cdot & \cdot & \dots & \cdot \\ \beta_{0n} & \beta_{1n} & \beta_{2n} & \dots & E \end{pmatrix}.$$

are determined uniquely by the blocks of $A$. $E \in C_p$ and $O \in C_p$ denote the unit and the zero matrix, respectively.

Let

$$x = (x_o, x_1, \dots, x_n),$$

$$x_j \in C_p \qquad (j=0,1,\dots,n)$$

be an arbitrary matrix-valued vector. By (1.9) the matrix-valued quadratic form $xAx^* \in C_p(H)$ can be written in the form

$$xAx^* = \xi_o \gamma_o \xi_o^* + \xi_1 \gamma_1 \xi_1^* + \dots + \xi_n \gamma_n \xi_n^* , \tag{1.10}$$

where

$$\xi = (\xi_o, \xi_1, \dots, \xi_n) = xB . \tag{1.11}$$

By using Definitions 1.1., 1.2. and 1.3. respectively, we can consider the following functionals of the matrix-valued quadratic form $xAx^*$ .

$$A(xAx^*) = A(A(\xi_j \gamma_j \xi_j^*) \quad (j=0,1,\dots,n)) ,$$

$$G(xAx^*) = G(G(\xi_j \gamma_j \xi_j^*) \quad (j=0,1,\dots,n)) ,$$

$$H(xAx^*) = H(H(\xi_j \gamma_j \xi_j^*) \quad (j=0,1,\dots,n)) .$$

Let us introduce the following notations:

$$A(A) = \inf A(xAx^*) , \tag{1.12}$$

if $\xi_j$ runs over $D$ $(j=0,1,\dots,n)$ ;

$$\underline{G}(A) = \inf G(xAx^*) , \tag{1.13}$$

if $\xi_j$ runs over $D$ $(j=0,1,\dots,n)$ ;

$$\overline{G}(A) = \sup G(xAx^*) , \tag{1.14}$$

if $\xi_j U_j$ runs over $P$ $(j=0,1,\dots,n)$ ;

$$H(A) = \sup H(xAx^*) , \tag{1.15}$$

if $\xi_j U_j$ runs over $P$ $(j=0,1,\dots,n)$ , where matrices $\xi_j$ $(j=0,1,\dots,n)$ are defined by (1.11), and

$$\gamma_j = U_j \Lambda_j U_j^* \in C_p^+(H) \quad (j=0,1,\dots,n) .$$

By Theorem 1.2. we have the following statement:

<u>Theorem 1.4.</u> Let the $(n+1)x(n+1)$ block matrix $A$ with

pxp matrices as its entries be Hermite symmetric and positive definite. Let $xAx^*\in C_p(H)$ be the corresponding matrix-valued quadratic form. Then equalities

$$A(xAx^*) = A(A) \ , \qquad H(xAx^*) = H(A)$$

are statisfied by the matrices $\xi_j$ $(j=0,1,\dots,n)$ in the case only if

$$\xi_j^*\xi_j = (\mathrm{Det}\ \gamma_j)^{1/p}\ \gamma_j^{-1} \qquad (j=0,1,\dots,n)\ .$$

Moreover equations

$$G(xAx^*) = \underline{G}(A) \ , \qquad G(xAx^*) = \overline{G}(A)$$

have the only solution $\xi_j$ $(j=0,1,\dots,n)$ satisfying conditions

$$\xi_j^*\xi_j = U_j d_j U_j^* \qquad (j=0,1,\dots,n) \ ,$$

where $d_j\in C_p^+(H)$ is a diagonal matrix with diagonal elements having product one. In this cases

$$A(A) = A((\mathrm{Det}\ \gamma_j)^{1/p} \quad (j=0,1,\dots,n)) \ ,$$

$$H(A) = H((\mathrm{Det}\ \gamma_j)^{1/p} \quad (j=0,1,\dots,n)) \ ,$$

$$\underline{G}(A) = \overline{G}(A) = (\mathrm{Det}\ A)^{1/(n+1)p} \ .$$

Obviously the following inequality holds too:

Theorem 1.5.

$$A(A) \geqq \underline{G}(A) = \overline{G}(A) \geqq H(A)$$

with equality if and only if

$$\mathrm{Det}\ \gamma_0 = \mathrm{Det}\ \gamma_1 = \dots = \mathrm{Det}\ \gamma_n \ .$$

## II. TOEPLITZ MATRICES

Denote by $Z$ the set of the non negative integers.

Let $M \in C_p$ be a matrix-valued measure defined on the circle such that $M(B)$ is Hermitian and positive semidefinite for every Borel set $B$. Let $M$ have Lebesgue decomposition

$$d\,M(e^{ix}) = W(e^{ix})d\sigma + d\,M_s(e^{ix}) \ , \quad d\sigma = \frac{dx}{2\pi} \ ,$$

where $W \in C_p(H)$ is a summable matrix-valued function, and $M_s \in C_p$ is singular with respect to $d\sigma$. The matrices

$$T_n(M) = T_n = (a_{jk})_o^n \ , \qquad n \in Z$$

are said to be Toeplitz matrices generated by the matrix-valued measure $M$, if

$$a_{jk} = \int_{-\pi}^{\pi} e^{i(j-k)x}\, d\,M(e^{ix}) \qquad (j,\ k \in Z) \ .$$

It is easy to verify that matrices $T_n$ are Hermitian, and positive semidefinite. Applying the generalized Lagrenge transformation, we have the representation (1.9) in the case where $T_n$ is a positive definite matrix, where $\gamma_j \in C_p^+(H)$, and the sequence $\{\gamma_n\}_o^\infty$ is monotonically nonincreasing in the sense of Löwner ([4]) such that

$$\lim_{n\to\infty} \gamma_n = \gamma$$

exists ([1]) Lemma 3, Corollary 1).

The sequence $\{\gamma_n\}_o^\infty$ is said to be the adjoined Lagrange sequence of $\{T_n\}_o^\infty$.

By Lemma 4 of [1], and by Theorem 8 of [3] we get the following Theorem.

<u>Theorem 2.1.</u> Let $\{T_n\}_o^\infty$ be the sequence of Toeplitz matrices generated by the matrix-valued measure $M \in C_p$ defined on the circle such that $M(B)$ is Hermitian and positive

semidefinite for every Borel set $B$ . Let $M$ have Lebesgue decomposition

$$d\,M(e^{ix}) = W(e^{ix})d\sigma + d\,M_s(e^{ix}) \;, \quad d\sigma = \frac{dx}{2\pi} \;,$$

where $W\in C_p(H)$ is a summable matrix-valued function and $M_s\in C_p$ is singular with respect to $d\sigma$ . Let $\{\gamma_n\}_o^\infty$ be the adjoint Lagrange sequence of $\{T_n\}_o^\infty$ . Then

$$(\mathrm{Det}\,\gamma_n)^{1/2} \searrow \exp\{\frac{1}{p}\int_{-\pi}^{\pi} \mathrm{tr}\,\lg W\,d\sigma\} \;, \quad n\to\infty \;,$$

The right hand side of the last formula is to be interpreted as zero if

$$\int_{-\pi}^{\pi} \mathrm{tr}\,\lg W\,d\sigma = -\infty.$$

By Theorem 1.4. and 2.1. we obtain the following result:

<u>Theorem 2.2.</u> Under the assumptions of Theorem 2.1. we get that

$$\lim_{n\to\infty} A(T_n) = \lim_{m\to\infty} \underline{G}(T_n) = \lim_{n\to\infty} \overline{G}(T_n) = \lim_{n\to\infty} H(T_n) =$$

$$= \exp\{\frac{1}{p}\int_{-\pi}^{\pi} \mathrm{tr}\,\lg W\,d\sigma\} \;,$$

where the left hand side of the last formula is to be interpreted as zero if

$$\int_{-\pi}^{\pi} \mathrm{tr}\,\lg W\,d\sigma = -\infty \;.$$

## III. THE EXTREME LINEAR PREDICTIONS OF THE MATRIX-VALUED STATIONARY STOCHASTIC SEQUENCES

By a matrix-valued random variable $X\in C_p$ defined on the probability space $(\Omega,F,P)$ we mean a quadratic matrix of order $p$ , the entries of which are random variables defined on this probability space.

By the expectation of $X \in C_p$ we mean the matrix $M(X) \in C_p$ formed by the expectation of the entries (assuming they exist). Let

$$E_p = \{X \in C_p \mid \exists M(X) \ , \quad \text{and} \quad \operatorname{tr} M(XX^*) < \infty \} \ .$$

The sequence

$$X = \{X_t \in E_p \ , \quad t = 0, \pm 1, \ldots\}$$

is said to be a matrix-valued weak stationary stochastic sequence if the conditions

$$M(X_t) = 0 \ , \quad M(X_t X_s^*) = R(t-s) \quad (t,s=0,\pm 1,\ldots) \tag{3.1}$$

are satisfied.

The matrix-valued function $R(t) \in C_p$ , $t = 0, \pm 1, \ldots$ is the covariance function of the matrix-valued stochastic sequence $X$ . For $R(t)$ we have also the analogy of the Herglotz representation, well-known for $p = 1$ . In fact ([2], p. 256)

$$R(t) = \frac{1}{2\pi} \int_{-\pi}^{\pi} e^{it\lambda} \, dF(\lambda) \ , \quad t = 0, \pm 1, \ldots,$$

where $F(\lambda) \in C_p(H)$ is a matrix-valued function defined on $[-\pi, \pi]$ , uniquely determined, non-decreasing in the sense of Löwner ([4]), of a bounded variation in its elements. $F(\lambda)$ is the matrix-valued spectral distribution function of the stochastic sequence $X$ .

Let us consider the linear form

$$\alpha_o X_o + \alpha_1 X_{-1} + \ldots + \alpha_n X_{-n} \tag{3.2}$$

of the random variables $X_o, X_{-1}, \ldots, X_{-n}$ , where

$$\alpha_j \in C_p \qquad (j=0,1,\ldots,n) \ .$$

By (3.1) and (3.2) it is easy to see that

$$M[(aX_1 - \sum_{k=0}^{n} \alpha_k X_{-k})^*(aX_1 - \sum_{k=0}^{n} \alpha_k X_{-k})] =$$

$$= \sum_{j=0}^{n} \sum_{k=0}^{n} R(j-k)u_j \bar{u}_k = u R_n u^* ,$$

where

$$\left.\begin{aligned} u_o = a,\ u_j = -\alpha_{j-1} \qquad (j = 1,\ldots,n) , \\ u = (u_o, u_1, \ldots, u_n) , \end{aligned}\right\} \tag{3.3}$$

moreover

$$R_n = (R(j-k))_{j,k=0}^{n} = T_n(F)$$

$$(n = 0,1,\ldots) .$$

We say that the linear form (3.2) of the random variables $X_o, X_{-1}, \ldots, X_{-n}$ is the extremal linear prediction of the random variable $X_1$ , if one of the equalities

$$A(u R_n u^*) = A(R_n) ,$$

$$G(u R_n u^*) = \underline{G}(R_n) ,$$

$$G(u R_n u^*) = \overline{G}(R_n) ,$$

$$H(u R_n u^*) = H(R_n)$$

is satisfied by the matrix-valued vector $u$ defined by (3.3), where the right hand side quantities of the equations are defined by (1.12), (1.13), (1.14) and (1.15), respectively. Quantities $A(R_n)$ , $\underline{G}(R_n)$ , $\overline{G}(R_n)$ , and $H(R_n)$ give us the measure of the errors of these linear predictions.

<u>Theorem 3.1.</u> Let

$$dF(\lambda) = f(\lambda)d\sigma + dF_S(\lambda)$$

be the Lebesgue decomposition of the matrix-valued spectral distribution $F(\lambda)$ of the matrix-valued stationary stochastic sequence $\{X_t \in E_p,\ t=0,\pm 1,\ldots\}$ . Suppose that

$$\int_{-\pi}^{\pi} \operatorname{tr} \lg f(\lambda)\, d\sigma > -\infty . \tag{3.4}$$

Let $\{\gamma_j\}_0^\infty$ be the adjoint Lagrange sequence of the sequence of Toeplitz matrices $\{R_n\}_0^\infty$ , and let

$$\gamma_j = U_j \Lambda_j U_j^* \in C_p^+(H) \ , \quad U_j U_j^* = E \ , \quad j \in Z \ .$$

Then the extremal linear matrix-valued prediction of $X_1$ by $X_0, X_{-1}, \ldots, X_{-n}$ is given by matrices $\xi_j \in C_p$ $(j=0,1,\ldots,n)$ for which either the conditions

$$\xi_j^* \xi_j = (\operatorname{Det} \gamma_j)^{1/p}\, \gamma_j^{-1} \qquad (j=0,1,\ldots,n) \ , \tag{3.5}$$

or conditions

$$\xi_j^* \xi_j = U_j\, d_j\, U_j^* \qquad (j=0,1,\ldots,n) \tag{3.6}$$

are satisfied, where $d_j \in C_p^+(H)$ is a diagonal matrix with diagonal elements having product one.

In the case (3.5) the measure of the error of linear prediction is given either by

$$A(R_n) = A((\operatorname{Det} \gamma_j)^{1/p} \quad (j=0,1,\ldots,n)) \ ,$$

or by

$$H(R_n) = H((\operatorname{Det} \gamma_j)^{1/p} \quad (j=0,1,\ldots,n)) \ ,$$

accordingly we use either functional $A$ or $H$ . In the case (3.6) the measure of the error of the linear prediction is given by

$$\underline{G}(R_n) = \overline{G}(R_n) = (\operatorname{Det} R_n)^{1/(n+1)p} \ .$$

<u>Proof.</u> By condition (3.4) all elements of the sequence

$\{R_n\}_0^\infty$ are Hermite symmetric positive definite matrices, hence Theorem 1.4. is applicable.

By Theorem 2.2. we get the following statement:

Theorem 3.2. Under the assumptions of Theorem 3.1. we get that

$$\lim_{n\to\infty} A(R_n) = \lim_{n\to\infty} H(R_n) = \lim_{n\to\infty} \underline{G}(R_n) = \lim_{n\to\infty} \overline{G}(R_n) =$$

$$= \exp\{\frac{1}{p}\int_{-\pi}^{\pi} \operatorname{tr} \lg f(\lambda)\, d\sigma\} \ .$$

## REFERENCES


[1] Gyires, B. 'A generalization of a theorem of Szegő'. Publ. of the Math. Inst. of the Hung. Acad. of Sciences, Vol. VIII. Ser. A. (1962). 43-51.

[2] Gyires, B. 'On the uncertainty of matrix-valued predictions'. Proc of the coll. on information theory. Edited by A. Rényi. János Bolyai Math. Soc. Budapest, 1968.

[3] Helson, H. and Lowdenslager, D. 'Prediction theory and Fourier series in several variables'. Acta Math. 99 (1958) 145-158.

[4] Löwner, K. 'Über monotone Matrixfunktionen'. Math. Zeitschr. 38 (1934), 177-216.

[5] Marcus, M. 'The permanent analogue of the Hadamard determinant theorem'. Bull. Amer. Math. Soc. 69 (1963), 494-496.

[6] Marcus, M.'The Hadamard theorem for permanents'. Proc. Amer. Math. Soc. 15 (1964), 967-973.

[7] Szász, O. 'Über die Verallgemeinerung des Hadamardschen Determinantensatzes'. Monatsh. f. Math. u. Phys. 28 (1917), 253-257.

# MULTIVARIATE B - SPLINES, ANALYSIS OF CONTINGENCY TABLES AND SERIAL CORRELATION

Z. G. Ignatov and V. K. Kaishev
Institute of Mathematics,
Bulgarian Academy of Sciences
1090 Sofia P.O. Box 373
Bulgaria

ABSTRACT. Numerically convenient divided difference formulae for the distribution function and the density of a linear combination of Dirichlet distributed random variables with all parameters integer except one, real are derived. It is shown how these formulae relate to B-splines and could be applied to compute the distribution of serial correlation coefficients and certain test statistics arising in the Bayesian analysis of contingency tables.

## 1. INTRODUCTION

The layout of the paper is as follows. In the next section we give a short discussion of B-splines and divided differences and review some of their important properties which will be used later. This background material will help readers not familiar with the theory of spline functions and the calculus of divided differences to readily follow the presentation in Sections 3 and 4. Further we elaborate a little on linear combinations of random variables having a joint Dirichlet distribution and their relation to B-splines. In particular, the interpretation of the density of such a linear combination as a multivariate B-spline, recently established by the authors of this note is given (see Lemma 2.1). Using it, the distribution of a linear combination of random variables having a joint Dirichlet distribution with all parameters save one positive integers the one - positive real is related to divided differences (Theorem 3.1, Section 3 ). This distributional problem was considered by Bloch and Watson(1967) and also by B. Margolin (1977). The corresponding formulae, given in both works are rather complex being characterized by the authors as 'too cumbersome for practical use'. In Section 3 (see Remark 1) we have given B. Margolin's formula. Theorem 3.1, Section 3 establishes a numerically convenient representation of the probability 1-F(x) as a divided difference of a certain function. Here F(x) denotes the distribution function of the corresponding linear combination. This representation can be directly



*P. Bauer et al. (eds.), Mathematical Statistics and Probability Theory, Vol. B, 125–137.*

used to compute recurrently significance points of linear combinations of Dirichlet distributed random variables.

Section 4 represents a discussion of some important applications of Lemma 2.1 and Theorem 3.1 to such areas of statistics as serial correlation and analysis of contingency tables.

## 2. PRELIMINARIES

In this section we give a collection of some basic properties of B-splines and linear combinations of Dirichlet distributed random variables which will be used further.

The (closed) convex hull of any set $A \subset R^s$ is denoted by $[A]$ and by $x, y, \ldots$ we will denote elements of the Euclidean space $R^s$.

Multivariate B-splines were introduced by C. de Boor (1976) in the following way. Let $\sigma = [y_0, \ldots, y_n]$ be any n-simplex in $R^n$, such that $y_i\big|_{R^s} = t_i$, $i=0,1,\ldots,n$ i.e. the first s coordinates of $y_i$ agree with the vector $t_i \in R^s$. The multivariate B-spline $M(t;t_0,\ldots,t_n)$ is defined as

$$M(t;t_0,\ldots,t_n) := \mathrm{vol}_{n-s}(\{u\in\sigma : u\big|_{R^s} = t\})/\mathrm{vol}_n(\sigma), \quad t \in R^s, \tag{2.1}$$

where $\mathrm{vol}_n(\sigma)$ is the n-dimensional volume of $\sigma$ and $t_0,\ldots,t_n \in R^s$ are called knots.

Obviously, $M(t;t_0,\ldots,t_n)$ is a nonnegative, locally supported function. More precisely, it is a piecewise polynomial of total degree not exceeding n-s, with n-s-1 continuous derivatives when the convex hull of every subset of s+1 points of $\{t_0,\ldots,t_n\}$ forms an s-dimensional simplex (see Micchelli 1980).

Of course, the characterization (2.1) can hardly serve as a practical representation of the B-spline. A numerically useful recurrent formula, relating higher order B-splines to lower order ones was found by C.A. Micchelli (1980). Thus, if $t = \sum_{j=0}^{n} \mu_j t_j$, $\sum_{j=0}^{n} \mu_j = 1$, then for $n > s+1$

$$M(t;t_0,\ldots,t_n) = \frac{n}{n-s} \sum_{j=0}^{n} \mu_j\, M(t;t_0,\ldots,t_{j-1},t_{j+1},\ldots,t_n). \tag{2.2}$$

For an exhaustive survey on multivariate B-splines, we refer to Dahmen and Micchelli (1983).

Definition (2.1) appeared as a natural generalization of the geometrical interpretation of the univariate B-spline given by Curry and Schoenberg (1966). They introduced the univariate B-spline as a divided difference of a certain function. The n-th order divided difference of any sufficiently smooth function $\varphi(u)$ over the points $t_0,\ldots,t_n \in R^1$ is defined as

$$[t_0,\ldots,t_n]_u\varphi(u)=\frac{[t_1,\ldots,t_n]_u\varphi(u) - [t_0,\ldots,t_{n-1}]_u\varphi(u)}{t_n - t_0}, \quad (2.3)$$

provided $t_0 \neq t_n$. If $t_0=t_1=\ldots=t_n$, then $[t_0,\ldots,t_n]_u\varphi(u) = D^n\varphi(t_0)/n!$ , where $D^n\varphi(t_0)$ is the n-th derivative of $\varphi(u)$ at $u=t_0$ , $n \geq 0$ , $(D^0\varphi(t_0):= \varphi(t_0))$. It is known (see Berezin and Jidkov 1966) that if, $\{t_0,\ldots,t_n\} = \{\underbrace{\tau_0,\ldots,\tau_0}_{\nu_0}, \underbrace{\tau_1,\ldots,\tau_1}_{\nu_1},\ldots, \underbrace{\tau_l,\ldots,\tau_l}_{\nu_l}\}$

with $\tau_0 < \tau_1 <\ldots< \tau_l$ , $\nu_0 + \nu_1+\ldots+ \nu_l = n+1$ , then

$$[t_0,\ldots,t_n]_u\varphi(u) = \sum_{i=0}^{l} D^{\nu_i-1} h_i(\tau_i)/(\nu_i- 1)! \quad , \quad (2.4)$$

where $h_i(u) = \varphi(u)/\prod_{\substack{j=0\\ j\neq i}}^{l} (u - \tau_j)^{\nu_j}$ .

Now, following Curry and Schoenberg (1966), the univariate B-spline $M(t;t_0,\ldots,t_n)$ of degree n-1 with knots $t_0,\ldots,t_n \in R^1$ is the n-th divided difference of the function $\Phi(u) = n(u-t)_+^{n-1}$ , i.e.,

$$M(t;t_0,\ldots,t_n):=[t_0,\ldots,t_n]_u\Phi(u) \quad , \quad (2.5)$$

where $(z)_+ = \max\{0,z\}$ .

Definition (2.5) together with Taylor's formula with integral remainder readily imply that if $t_0 < t_n$ , $t_0 \leq t_i \leq t_n$, $i = 1,\ldots,n-1$, then

$$[t_0,\ldots,t_n]_u\varphi(u) = \int_{t_0}^{t_n} M(y;t_0,\ldots,t_n)\varphi^{(n)}(y)dy/n! \quad , \quad (2.6)$$

for all $\varphi(u) \in L_1^n(t_0,t_n)$. This is the well known (see e.g. Schumaker 1981 page 128) Peano representation for divided differences, which we will need to prove Theorem 3.1.

An efficient and stable means for computation of univariate B-splines is supplied by the de Boor-Cox (see e.g. de Boor 1976) recurrence expression

$$M(t;t_0,\dots,t_n) = (n/(n-1))(M(t;t_0,\dots,t_{n-1})(t - t_0) + $$

$$+ M(t;t_1,\dots,t_n)(t_n - t))/(t_n - t_0) \quad , \quad (2.7)$$

where $t_0,\dots,t_n \in R^1$, $n \geq 2$.

Now, consider linear combinations of Dirichlet distributed random variables. Recall that the random variables $\theta_0, \theta_1,\dots, \theta_n$ have the joint Dirichlét distribution $\mathcal{D}(g_0,g_1,\dots,g_n)$ with parameters $g_0 > 0$, $g_1 > 0,\dots,g_n > 0$, $((\theta_0,\theta_1,\dots,\theta_n) \in \mathcal{D}(g_0,g_1,\dots,g_n))$ if $\theta_0 = 1 - \theta_1 - \dots - \theta_n$, and the joint probability density of $\theta_1,\theta_2,\dots,\theta_n$ with respect to the Lebesgue measure on the simplex

$S^n = \{(u_1,\dots,u_n)\colon u_i \geq 0 \ , \ \sum_{i=1}^{n} u_i \leq 1 \}$ , $u_0 = 1 - u_1 - \dots - u_n$ is

$$\frac{\Gamma(g_0+\dots+ g_n)}{\Gamma(g_0)\dots\Gamma(g_n)} (1 - u_1 - u_2 - \dots - u_n)^{g_0 - 1} u_1^{g_1 - 1} \dots u_n^{g_n - 1}$$

($\Gamma(.)$ is the well known Gamma function).

The following result, recently found by Ignatov and Kaishev (1985), gives an important relation between B-splines and linear combinations of Dirichlet distributed random variables.

Lemma 2.1. Let $(\theta_0,\dots,\theta_n) \in \mathcal{D}(g_0,\dots,g_n)$, $g_i$ - positive integers, $i = 0,\dots,n$ and let $t_0,\dots,t_n$ be points in $R^s$. The probability density function of $\sum_{i=0}^{n} \theta_i t_i$ is the multivariate B-spline

$$M(t;\underbrace{t_0,\dots,t_0}_{g_0}, \underbrace{t_1,\dots,t_1}_{g_1}, \dots, \underbrace{t_n,\dots,t_n}_{g_n}) .$$

In the literature on spline functions the parameters $g_0,\dots,g_n$ are called multiplicities of the knots $t_0,\dots,t_n$. If we allow $g_i, i=0,\dots,n$ to take real values, then we arrive at the notion of a generalized B-spline, i.e., one with rational multiplicities of the knots. In the next section we consider a special case of such generalized B-splines.

Now, using Lemma 2.1 in the univariate case, i.e., when $s = 1$, we can express the Peano representation (2.6) as

Lemma 2.2. Let $(\theta_0,\dots,\theta_n) \in \mathcal{D}(g_0,\dots,g_n)$, $g_i$ - positive integer, $i = 0,\dots,n$, $d = g_0+\dots+ g_n$ and $\Psi(u) \in C^{(d-1)}([A])$, where the set

$A = \{t_0,\ldots,t_n\} \subset R^1$. Then the expectation

$$E\{\varphi^{(d-1)}(t_0\theta_0+\ldots+t_n\theta_n)/(d-1)!\}=[\underbrace{t_0,\ldots,t_0}_{g_0},\ldots,\underbrace{t_n,\ldots,t_n}_{g_n}]_u\varphi(u)$$

## 3. LINEAR COMBINATIONS OF DIRICHLET VARIABLES AND DIVIDED DIFFERENCES

For reasons given below, (see Remark 3) we will restrict our attention here to the univariate, i.e., when $s = 1$, generalized B-spline with only one knot having noninteger multiplicity. We study such a generalized B-splines by investigating the distribution of the linear combination of Dirichlet distributed random variables whose density this B-spline is. More precisely, our purpose will be to see how the latter distribution can de expressed in terms of a divided difference of a certain function. Such an expression seems to be numerically more appealing than the existing formulae of Bloch and Watson (1967) and B. Margolin (1977) (see Remark 1).

Let the random variables $\theta_0,\ldots,\theta_n$ have a joint Dirichlet distribution with parameters $\rho,g_1,\ldots,g_n$, i.e., $(\theta_0,\ldots,\theta_n) \in \mathcal{D}(\rho,g_1,\ldots,g_n)$, where $\rho$ - positive real, $g_i$ - positive integer, $i = 1,2,\ldots,n$. Let also $t_i$, $i = 0,1,\ldots,n$ be pairwise distinct real numbers and denote by $\hat{\rho}$ the integral part of $\rho$, $\bar{\rho} = \rho - \hat{\rho}$, $l = \hat{\rho} + g_1+\ldots+ g_n$.

Consider the random variable

$$S = t_0\theta_0+ t_1\theta_1+\ldots+ t_n\theta_n \ . \tag{3.1}$$

Theorem 3.1. The probability $P(S > x)$ is the $(l-1)$-th order divided difference in $u$ at the points $\underbrace{t_0,\ldots,t_0}_{\hat{\rho}}$, $\underbrace{t_1,\ldots,t_1}_{g_1},\ldots,$ $\underbrace{t_n,\ldots,t_n}_{g_n}$ of the function

$$G(u) = \frac{\Gamma(l + \bar{\rho})}{\Gamma(\bar{\rho})\Gamma(l)} \int_0^1 y^{\bar{\rho}-1} (u - x - (u - t_0)y)_+^{l-1}\, dy \quad , \text{ i.e.,}$$

$$P(S > x) = [\underbrace{t_0,\ldots,t_0}_{\hat{\rho}}, \underbrace{t_1,\ldots,t_1}_{g_1},\ldots,\underbrace{t_n,\ldots,t_n}_{g_n}]_u G(u) \ . \tag{3.2}$$

Proof. Let us first consider the case $\rho > 1$ . It can be directly verified that

$$P(S > x) = \int_{R^1} f_S(u)\, \Psi^{(l-1)}(u)\, du \quad , \tag{3.3}$$

where $\Psi^{(l-1)}(u)$ is the $(l-1)$-th derivative of the function $\Psi(u)=(u-x)_+^{l-1}/(l-1)!$ and $f_S(u)$ is the probability density function of S.

Obviously the right-hand side of (3.3) can be viewed as the expectation $E\Psi^{(l-1)}(S)$ . We can express the latter as

$$\begin{aligned} E\Psi^{(l-1)}(S) &= E\Psi^{(l-1)}(t_0\theta_0+\ldots+ t_n\theta_n) \\ &= E\Psi^{(l-1)}(t_0\theta_0' + t_0\theta_0'' + t_1\theta_1+\ldots+ t_n\theta_n) \quad , \end{aligned} \tag{3.4}$$

where $(\theta_0' , \theta_0'' , \theta_1,\ldots,\theta_n) \in \mathcal{D}(\bar{\rho}, \hat{\rho}, g_1,\ldots,g_n)$.

We have

$$E\Psi^{(l-1)}(t_0\theta_0' + t_0\theta_0'' + t_1\theta_1+\ldots+t_n\theta_n) \tag{3.5}$$

$$= E\Psi^{(l-1)}(t_0\theta_0' \frac{\theta_0'' + \theta_1+\ldots+ \theta_n}{1 - \theta_0'} + (1 - \theta_0')(t_0 \frac{\theta_0''}{1 - \theta_0'} + t_1 \frac{\theta_1}{1 - \theta_0'} + \ldots + t_n \frac{\theta_n}{1 - \theta_0'}))$$

$$= E\Psi^{(l-1)}(((t_0\theta_0' + (1 - \theta_0' )t_0) \frac{\theta_0''}{1 - \theta_0'} + (t_0\theta_0' + (1 - \theta_0')t_1)\frac{\theta_1}{1 - \theta_0'} +\ldots+(t_0\theta_0' + (1 - \theta_0')t_n)\frac{\theta_n}{1 - \theta_0'} )$$

$$= E(E\{\Psi^{(l-1)}((t_0\theta_0' + (1 - \theta_0')t_0)\frac{\theta_0''}{1 - \theta_0'} + (t_0\theta_0'+(1-\theta_0')t_1) \frac{\theta_1}{1 - \theta_0'} +\ldots+(t_0\theta_0'+(1-\theta_0')t_n)\frac{\theta_n}{1 - \theta_0'}) \mid \theta_0'\}).$$

In the last equation, $E\{.\mid\theta_0'\}$ is the conditional expectation given $\theta_0'$.

It is known (cf Wilks 1962 ,page 177) that the conditional

distribution of $\dfrac{\theta_0''}{1-\theta_0'}$, $\dfrac{\theta_1}{1-\theta_0'}$, ..., $\dfrac{\theta_n}{1-\theta_0'}$, given $\theta_0'$,

coincides with the Dirichlet distribution with parameters $\hat{\rho}, g_1, \ldots, g_n$.

From (3.3), (3.4), (3.5) and Lemma 2.2, we have

$$P(S>x)=(1-1)!E[\underbrace{\tilde{t}_0,\ldots,\tilde{t}_0}_{\hat{\rho}},\underbrace{\tilde{t}_1,\ldots,\tilde{t}_1}_{g_1},\ldots,\underbrace{\tilde{t}_n,\ldots,\tilde{t}_n}_{g_n}]_u\Psi(u), \quad (3.6)$$

where $\tilde{t}_i = t_0\theta_0' + (1-\theta_0')t_i$, $i = 0, 1, \ldots, n$.

Now, applying formula (2.4) to the divided difference in the right-hand side of (3.6) we obtain

$$P(S>x)=(1-1)!E\left\{\sum_{i=0}^{n} D_u^{g_i-1} \frac{(u-x)_+^{1-1}}{((1-1)!)T(i,u)}\Bigg|_{u=\tilde{t}_i}\right\}, \quad (3.7)$$

where we have set $g_0 = \hat{\rho}$ and $T(i,u) = \prod_{\substack{j=0\\ j\neq i}}^{n} (u-t_j)^{g_j}$.

It can be directly verified that

$$\sum_{i=0}^{n} D_u^{g_i-1} \frac{(u-x)_+^{1-1}}{T(i,u)}\Bigg|_{u=\tilde{t}_i} = \sum_{i=0}^{n} D_u^{g_i-1} \frac{((u-(x-t_0\theta_0')(1-\theta_0')^{-1})_+^{1-1}}{T(i,u)}\Bigg|_{u=\tilde{t}_i}.$$

Hence,

$$P(S>x) = E\sum_{i=0}^{n} D_u^{g_i-1} \frac{((u-(x-t_0\theta_0')(1-\theta_0')^{-1})_+^{1-1}}{T(i,u)}\Bigg|_{u=\tilde{t}_i} \quad (3.8)$$

$$= E[\underbrace{t_0,\ldots,t_0}_{g_0},\ldots,\underbrace{t_n,\ldots,t_n}_{g_n}]_u((u-(x-t_0\theta_0')(1-\theta_0')^{-1})_+^{1-1}.$$

Now changing the order of integration and divided difference in (3.8), we obtain

$$P(S>x) = E[\underbrace{t_0,\ldots,t_0}_{g_0},\ldots,\underbrace{t_n,\ldots,t_n}_{g_n}]_u E((u-(x-t_0\theta_0')(1-\theta_0')^{-1})_+^{1-1}.$$

Since the density of $\theta_0'$ is $f_{\theta_0'}(y) = \dfrac{\Gamma(1+\bar{\rho})}{\Gamma(1)\Gamma(\bar{\rho})} y^{\bar{\rho}-1}(1-y)^{1-1}$, $0 < y < 1$

and $f_{\theta_0'}(y) = 0$ otherwise, then

$$E((u-(x-t_0\theta_0')(1-\theta_0')^{-1})_+^{l-1}$$

$$= \frac{\Gamma(l+\bar{\rho})}{\Gamma(l)\Gamma(\bar{\rho})} \int_0^1 y^{\bar{\rho}-1}(1-y)^{l-1}((u-(x-t_0y)(1-y)^{-1})_+^{l-1}$$

$$= \frac{\Gamma(l+\bar{\rho})}{\Gamma(l)\Gamma(\bar{\rho})} \int_0^1 y^{\bar{\rho}-1}(u-x-(u-t_0)y)_+^{l-1}\, dy \quad ,$$

and we obtained the assertion of the Theorem for the case $\rho > 1$.

When $0 < \rho < 1$ the proof is virtually the same with the role of $\theta_0'$ taken by $\theta_0$, and $\theta_0''$ deleted. In this case, $\hat{\rho} = 0$, and

$$[\underbrace{t_0,\ldots,t_0}_{\hat{\rho}},\ \underbrace{t_1,\ldots,t_1}_{g_1},\ldots,\underbrace{t_n,\ldots,t_n}_{g_n}]_u\Psi(u)$$

$$:= [\underbrace{t_1,\ldots,t_1}_{g_1},\ldots,\underbrace{t_n,\ldots,t_n}_{g_n}]_u\Psi(u) \ .$$

This completes the proof of the Theorem.

Remark 1. Different expression for $P(S > x)$ from Theorem 3.1 in the special case, $t_0 = 0$ and $t_1 > t_2 > \ldots > t_n$ was given by B. Margolin (1977), i.e.,

$$P(S > x) = \begin{cases} \sum\limits_{j\in J_+} \sum\limits_{k=1}^{g_j} \alpha_{jk} \sum\limits_{i=0}^{k-1} \dfrac{\Gamma(d)\ \zeta_j^i\ (1-\zeta_j)^{d-i-1}}{\Gamma(d-i)\Gamma(i+1)} \ , & x > 0 \\ 1 - \sum\limits_{j\in J_-} \sum\limits_{k=1}^{g_j} \alpha_{jk} \sum\limits_{i=0}^{k-1} \dfrac{\Gamma(d)\ \zeta_j^i\ (1-\zeta_j)^{d-i-1}}{\Gamma(d-i)\Gamma(i+1)} \ , & x \le 0 \end{cases} \tag{3.9}$$

where $J_+ = \{j:t_j > \max(0,x)\}$, $J_- = \{j:t_j < \min(0,x)\}$, $\zeta_j = x/t_j$, $j = 1,\ldots,n$, $d = \rho + g_1+\ldots+ g_n$, and $\alpha_{jk}$ is $(g_j - k)!$ times the value at $y=0$ of the $(g_j- k)$-th derivative with respect to $y$ of the function

$$\prod_{i\neq j}^{n} ((t_j - t_i)/\ t_j + (yt_i)/t_j)^{-g_i} \ .$$

Corollary 1. If $t_0 < t_i$ , $i = 1,\dots,n$ , then

$$P(S > x) = \begin{cases} [\underbrace{t_0,\dots,t_0}_{\hat{\rho}},\ \underbrace{t_1,\dots,t_1}_{g_1},\dots,\underbrace{t_n,\dots,t_n}_{g_n}]_u H(u),\ t_0 \le x \\ 1 \qquad \text{otherise} , \end{cases} \tag{3.10}$$

where $H(u) = (u-t_0)^{-\bar{\rho}}(u-x)_+^{\rho+g_1+\dots+g_n-1}$ , $H(t_0) := 0$ .

Proof. From (3.2) and the condition $t_0 \le x$ if $u > t_0$, then we have

$$\frac{\Gamma(1+\bar{\rho})}{\Gamma(1)\Gamma(\bar{\rho})} \int_0^1 y^{\bar{\rho}-1}(u - x - (u - t_0)y)_+^{1-1}\, dy$$

$$= \frac{\Gamma(1+\bar{\rho})(u-t_0)^{1-1}}{\Gamma(1)\Gamma(\bar{\rho})} \int_0^{(w)_+} y^{\bar{\rho}-1}(w - y)^{1-1}\, dy = H(u) ,$$

where $w = (u-x)/(u-t_0)$ .

The last equality follows after integrating (1-1) times by parts.

Corollary 2. If $t_0 < t_i$, $i = 1,\dots,n$ then the probability density function of S is

$$f_S(x) = \begin{cases} [\underbrace{t_0,\dots,t_0}_{\hat{\rho}},\ \underbrace{t_1,\dots,t_1}_{g_1},\dots,\underbrace{t_n,\dots,t_n}_{g_n}]_u H'(u),\ x \in [A] \\ 0 \qquad \text{otherwise} , \end{cases} \tag{3.11}$$

where $H'(u) = (1+\bar{\rho}-1)(u-t_0)^{-\bar{\rho}}(u-x)_+^{1+\bar{\rho}-2}$ , $A = \{t_0,\dots,t_n\}$ .

Proof. Differentiating (3.10) in x we obtain (3.11).

Remark 2. If $\bar{\rho}=0$ then the divided difference in formulae (3.10) and (3.11) are equal respectively to 1 and 0, for every $x < t_0$. The same is not true when $\bar{\rho}$ is positive.

Remark 3. If we consider a linear combination of Dirichlet distributed random variables with two or more parameters - positive real then the distribution function of the random variable S does not admit a representation as a divided difference of an elementary function as in Corollaries 1 and 2. For example, if $(\theta_0,\theta_1) \in \mathcal{D}(1/2,\ 1/2)$ then $S = 0\theta_0 + 1\theta_1 = \theta_1$ has a distribution function

$$P(S > x) = P(\theta_1 > x) = \frac{\Gamma(1/2 + 1/2)}{\Gamma(1/2)\Gamma(1/2)} \int_x^1 ((1-u)u)^{-1/2}\, du$$

$$= 1/2 - (1/\pi)\arcsin(2x - 1)$$

which is a transcendental function.

## 4. APPLICATIONS

In this section, attention will be turned to some interesting applications of the probabilistic interpretation of B-splines, given by Lemma 2.1, Section 2 and also of Theorem 3.1, Section 3. On basis of these results, we will show that B-splines and divided differences could be useful in dealing with serial correlation coefficients and contingency tables.

### 4.1. Bayesian analysis of contingency tables

Let $\xi_0, \xi_1, \dots, \xi_n$ be a sample from the multinomial distribution with cell probabilities $\theta_0, \theta_1, \dots, \theta_n$ respectively, $\sum_{i=0}^{n} \theta_i = 1$. In particular, let $\xi_0, \xi_1, \dots, \xi_n$ be data from a contingency table. Using the Bayesian aproach to the statistical analysis of the unknown parameters $\theta_0, \theta_1, \dots, \theta_n$, Lindley (1964) studies the case when the prior distribution of $(\theta_0, \theta_1, \dots, \theta_n)$ is Dirichlet, i.e., $(\theta_0, \theta_1, \dots, \theta_n) \in \mathcal{D}(1,1,\dots,1)$. Then the posterior distribution of $(\theta_0, \theta_1, \dots, \theta_n)$ given $\xi_0, \xi_1, \dots, \xi_n$ is also a Dirichlet distribution with parameters $1 + \xi_0, 1 + \xi_1, \dots, 1 + \xi_n$, i.e.,

$$\mathcal{L}(\theta_0, \theta_1, \dots, \theta_n \mid \xi_0, \xi_1, \dots, \xi_n) \in \mathcal{D}(1+\xi_0, 1+\xi_1, \dots, 1+\xi_n)$$

(see Bloch and Watson 1967, Section 4).

In many problems, related to analysis of contingency tables, linear combinations of $\theta_0, \theta_1, \dots, \theta_n$ appear as test statistics. For examples, we refer to the genetic linkage problem, described by Fisher (1928), Chapter 9 and treated by Bloch and Watson (1967) as well as to some simpler problems of this kind, considered by Lindley (1965), Chapter 7.

Lemma 2.1, combined with formulae (2.2) or (2.7), Section 2 give a convenient means for computation of various critical regions, significant points etc. of the corresponding test statistics which represent linear combinations of $\theta_0, \theta_1, \dots, \theta_n$.

### 4.2. Serial correlation coefficients

The serial correlation coefficient (SCC) has been introduced (see J. von Neumann 1941, R.L. Anderson 1942, T. Koopmans 1942, P. Hsu 1946) as a measure of the relationship between successive values of a variable, ordered in time or space. Serial correlation coefficients have important applications to regression problems (see Watson and Durbin 1951), to hypothesis testing in analysis of time series (see G. Jenkins 1956, T. W. Anderson 1971).

Let $y_1, y_2, \ldots, y_N$ be variables representing a random sample of N successive observations from a population whose distribution is normal with mean $\mu$ and variance $\sigma^2$. The lag j circular serial correlation coefficient in the case of unknown population mean is defined as

$$r_j = \sum_{i=1}^{N} (y_i - \bar{y})(y_{i+j} - \bar{y}) / \sum_{i=1}^{N} (y_i - \bar{y})^2 \quad , \tag{4.1}$$

where $y_{N+i} = y_i$ , $\bar{y} = (y_1 + \ldots + y_N)/N$ and of zero mean as

$$r_j = \sum_{i=1}^{N} y_i y_{i+j} / \sum_{i=1}^{N} y_i^2 \quad , \; y_{N+i} = y_i \; , \quad j < N \; . \tag{4.2}$$

To test for serial correlation the following noncircular test statistics is also available

$$r_j = \sum_{i=1}^{N-j} (\Delta^j y_i)^2 / \sum_{i=1}^{N} y_i^2 \quad , \qquad j < N \tag{4.3}$$

where $\Delta^j$ , the successive difference operator of order j, is given as

$$\Delta^j y_i = \sum_{v=0}^{j} (-1)^{j-v} \binom{j}{v} y_{i+v}$$

The marginal distribution of $r_j$ in (4.1) has been given by R. Anderson (1942), while the joint distribution of $r_1, \ldots, r_p$, $p < N$ is derived by M. Quenouille (1949) and by Watson (1956). The SCC in (4.2) and (4.3) were studied by T.W. Anderson (1971).

It is known (see e.g., B. Margolin 1977) that the distribution of each SCC in (4.1), (4.2) and (4.3) coincides with the distribution of the linear combination of Dirichlet distributed random variables (3.1) with $\rho = 1/2$ and $g_i$, $t_i$ suitable constants. Hence, we can use Theorem 3.1 to compute recurrently significant points of the distribution of each SCC in (4.1), (4.2) and (4.3).

Let us note (see Ignatov and Kaishev 1986) that the density of the joint distribution of $r_1, \ldots, r_p$ , $p < N$ from (4.1) when N is odd $(N = 2k + 1)$ coincides with the multivariate B-spline $M(t; t_0, \ldots, t_n)$ where $t \in R^p$ , $n = k - 1$ and

$$t_i = (\cos(2\pi(i+1)/N) \; , \; \cos(2\pi 2(i+1)/N), \ldots, \; \cos(2\pi p(i+1)/N) \; ,$$

$i = 0, 1, \ldots, k-1$ . In this case, the recurrence formula (2.2) can be successfully applied for computations involving the joint density of $r_1, \dots, r_p$ .

References

Anderson, R.L. (1942). 'Distribution of the serial correlation coefficient'. Ann. Math. Statist. 13, 1-13.

Anderson, T.W. (1971). The statistical analysis of time series, J. Wiley, New York.

Berezin, I.S. and Jidkov, N. P. (1966). Numerical Methods. 1 Nauka, Moskow, (in Russian).

Bloch, D.A. and Watson, G.S. (1967). 'A Bayesian study of the multinomial distribution.' Ann. Math. Statist. 38, 1423-35.

de Boor, C. (1976). 'Splines as linear combinations of B-splines' in: Approximation Theory II, ed. by G.G. Lorentz, C.K. Chui and L.L. Schumaker, Academic Press, New York, 1-47.

Curry, H.B. and Schoenberg, I.J. (1966). 'On Polya frequency functions. IV. The fundamental spline functions and their limits' J. d'Analyse Math. 17, 71-107.

Dahmen, W. and Micchelli, C.A. (1983). 'Recent progress in multivariate splines' in: Approximation Theory IV, Acad. Press, New York, ed. by C.K. Chui, L.L. Schumaker, T.D. Ward

Fisher, R.A. (1928). Statistical Methods for Research Workers, second and later editions, Hafner, New York.

Hsu, P.L. (1946). 'On the asymptotic distribution of certain statistics used in testing the independence between successive observations from a normal population'. Ann. Math. Statist. 13, 14-33.

Ignatov, Z.G. and Kaishev, V.K. (1985). 'A probabilistic interpretation of multivariate B-splines and some applications', to appear.

Ignatov, Z.G. and Kaishev, V.K. (1986). 'B-splines and serial correlation' in: Proceedings of the IV-th international conference on statistical methods in experimental research and quality control, 1 Varna.

Jenkins, G.M. (1956). 'Test of hypothesis in the linear autoregressive model'. II. Biometrika, **43**, 186-99.

Koopmans, T. (1942). 'Serial correlation and quadratic forms in normal variables'. Ann. Math. Statist. **13**, 14-33.

Lindley, D.V. (1964). 'The Bayesian analysis of contingency tables'. Ann. Math. Statist. **35**, 1622-43.

Lindley, D.V. (1965). Probability and Statistics from a Bayesian point of view, **2** Cambridge University Press, England

Margolin, B.H. (1977). 'The distribution of internally studentized statistics via Laplace transform inversion'. Biometrika, **64**, 3, 573-82.

Micchelli, C. A. (1980). 'A constructive approach to Kergin interpolation in $R^k$: Multivariate B-splines and Lagrange interpolation'. Rocky Mountain J. Math. **10**, 3, 485-497.

von Neumann, J. (1941). 'Distribution of the ratio of the mean square successive difference to the variance'. Ann. Math. Statist. **12**, 367-95.

Quenouille, M.H. (1949). 'The joint distribution of serial correlation coefficients'. Ann. Math. Statist. **20**, 561-71.

Schumaker, L.L. (1981). The theory of spline functions, J. Wiley, New York.

Watson, G.S. (1956). 'On the distribution of the circular serial correlation coefficients'. Biometrika, **43**, 161-68.

Watson, G.S. and Durbin, J. (1951). 'Exact tests of serial correlation using non-circular statistics'. Ann. Math. Statist. **22**, 446-51.

Wilks, S.S. (1962). Mathematical Statistics. J. Wiley, New York.

# DISCRIMINATION BETWEEN STATIONARY GAUSSIAN TIME SERIES

M. Krzyśko and J. Wachowiak
Adam Mickiewicz University
Institute of Mathematics
60-769 Poznań
Poland

ABSTRACT. In the present paper discrimination among multivariate Gaussian processes by Bayes method is described. Associated with any classification rule are the probabilities of misclassifying an observation. To calculate these probabilities we must know the distribution of the discriminant function. The exact distribution of the discriminant function is very complicated and not useful in practice. Now, the limit distribution of the discriminant function is investigated. The result is based on Lapunov's central limit theorem for series.

## 1. INTRODUCTION

A traditional classification problem arises in testing a decision procedure that involves simple observations. Usually the observations are taken at one time, but in many practical situations problems arise in which observations on one or more random variables are made on an individual at successive points in time (see e.g. Shumway, 1982). The goal is to classify the individual into one of two or more populations on the basis of these observations.

In the present paper discrimination among multivariate stationary Gaussian processes by Bayes method is described.

## 2. BAYES METHOD OF CLASSIFICATION

Let $\pi_i$, $i=1,2,\ldots,k$, denote a class of p-component vector stationary Gaussian processes $\{X_i(t),\ t=0,\pm1,\pm2,\ldots\}$ with expected value $E\{X_i(t)\} = u_i$ and covariance function $E\{(X_i(m)-\mu_i)(X_i(n)-\mu_i)'\} = \Gamma_i(n-m) = \Gamma_i'(m-n)$. Let us assume



*P. Bauer et al. (eds.), Mathematical Statistics and Probability Theory, Vol. B, 139–146.*

that we have a realization $X = (X(1), X(2), \dots, X(T))$ of a process belonging to one of the classes $\pi_1, \pi_2, \dots, \pi_k$.

The mathematical problem of discrimination is to divide the observation space $R^N$ ($N=pT$) into $k$ mutually exclusive and exhaustive regions $R_1, \dots, R_k$, so that if an observation $X$ falls in $R_i$, then it is classified as belonging to the $i^{th}$ population. We consider $k$ alternative hypotheses $H_1, H_2, \dots, H_k$ forming a complete system of disjoint events, i.e. if $q_i$ ($q_i > 0$), $i=1,2,\dots,k$, is the a priori probability of the acceptability of hypothesis $H_i$ then $q_1+q_2+\dots+q_k=1$. The hypothesis $H_i$ states that $X$ is a realization of a process belonging to class $\pi_i$, $i=1,2,\dots,k$. Our task is to verify the acceptability of one of these $k$ alternative hypothesis. In the statistical literature (see, for example Anderson, 1958, Chap. 6) this problem is known as classification problem. We will make use of the Bayesian method of classification dividing $R^N$ into non-intersecting classification regions $R_1, R_2, \dots, R_k$ defined so as to minimize the Bayesian risk

$$r = 1 - \sum_{i=1}^{k} q_i \int_{R_i} f_i(X)\,dX, \qquad (1)$$

where $f_i(X) = f_i(X(1), \dots, X(T))$ is the density function of the joint distribution of $X$ in class $\pi_i$:

$$f_i(X) = (2\pi)^{-\frac{N}{2}} |G_i|^{-\frac{1}{2}} \exp\left\{-\tfrac{1}{2}\mathrm{vec}'(X-m_i)G_i^{-1}\mathrm{vec}(X-m_i)\right\},$$

where

$m_i = \mu_i 1'$, $1' = (1,1,\dots,1)$, $i=1,2,\dots,k$, while $\mathrm{vec}\,A$ is a vector formed by stacking the columns of $A$, one on top of the other, in order from left to right. The matrix $G_i$ is a block matrix of $T^2$ blocks, where block $(u,v)$ contains the matrix $\Gamma_i(v-u)$, $i=1,2,\dots,k$.

The region $R_i$ minimizing the Bayesian risk (1) is

$$R_i = \left\{X:\ v_{ij}(X) \geqslant \ln(q_j/q_i),\ j=1,2,\dots,k,\ j \neq i\right\}, \qquad (2)$$

where $v_{ij}$ is a discriminant function of the following form

$$v_{ij}(X) = \ln(f_i(X)/f_j(X)) = \tag{3}$$
$$= -\frac{1}{2}(\mathrm{vec}'(X-m_i)G_i^{-1}\mathrm{vec}(X-m_i)-\mathrm{vec}'(X-m_j)G_j^{-1}\mathrm{vec}(X-m_j))+$$
$$+\frac{1}{2}\ln(|G_j|/|G_i|)), \quad i,j=1,2,\ldots,k,\ j\neq i.$$

According to the Bayesian method, $H_i$ is accepted if $X\in R_i$, $i=1,2,\ldots,k$.

When the hypothesis $H_i$ is true the probability $\alpha_i$ of rejecting it is equal to

$$\alpha_i=1-P(v_{ij}(X)\geq \ln(q_j/q_i),\ j=1,2,\ldots,k,\ j\neq i|H_i),$$
$$i=1,2,\ldots,k,$$

and the minimum value of the Bayesian risk is

$$r=1-\sum_{i=1}^{k} q_i P(v_{ij}(X)\geq \ln(q_j/q_i),\quad j=1,\ldots,k,\ j\neq i|H_i) .$$

The probability $\alpha_i$ is very difficult to evaluate in this case because of the very complicated shapes of the classification regions.

An alternative way of obtaining an upper bound for $\alpha_i$ is given by using the Bonferroni inequality which yields

$$\alpha_i \leq \sum_{\substack{j=1\\ j\neq i}}^{k} P(v_{ij}(X)< \ln(q_j/q_i)|H_i),\quad i=1,2,\ldots,k \tag{4}$$

and

$$r\leq \sum_{i=1}^{k}\sum_{\substack{j=1\\ j\neq i}}^{k} q_i P(v_{ij}(X)<\ln(q_j/q_i)|H_i). \tag{5}$$

If $k=2$ , then equality holds in the above formulas.
We see that to calculate an upper bound for the Bayesian risk we need to know the distribution of $v_{ij}(X)$, $i,j=1,2,\ldots,k,\ j\neq i$.
The exact distribution of this function is very complicated and not useful in practice, Krzyśko (1985).

We are now passing on to the asymptotic distribution of the discriminant function $v_{ij}(X)$.

## 3. ASYMPTOTIC DISTRIBUTION OF THE DISCRIMINANT FUNCTION

Since $G_i^{-1}$ is a symmetric p.d. matrix, there exists a nonsingular matrix $A_i$ such that $G_i^{-1}=A_i'A_i$, $i=1,2,\ldots,k$.
For $n=1,\ldots,N$ let $\lambda_{ij,n}$ ( $\lambda_{ij,1}\leq\cdots\leq\lambda_{ij,pT_0}\neq 1$,

$\lambda_{ij,pT_o+1} = \dots = \lambda_{ij,pT} = 1, pT_o = N_o)$ be the eigenvalue and $P_{ij,n}$ be the corresponding normalized eigenvector of $R_{ij} = (A_i')^{-1}G_j^{-1}A_i^{-1}$, and let $P_{ij} = (P_{ij,1},\dots,P_{ij,N})$ and $\lambda_{ij} = (\lambda_{ij,1},\dots,\lambda_{ij,N})$, where $i,j=1,2,\dots,k$, $j\neq i$.

Writing $\text{vec}(X-m_j)=\text{vec}(X-m_i)+\text{vec}(m_i-m_j)$ we have

$$v_{ij}(X) = -\frac{1}{2}\Big[(P_{ij}'A_i\text{vec}(X-m_i))'(I-\Lambda_{ij})(P_{ij}'A_i\text{vec}(X-m_i))- - \gamma_{ij}'(P_{ij}'A_i\text{vec}(X-m_i))\Big] + k_{ij},$$

where

$$\Lambda_{ij}=\text{diag}(\lambda_{ij,1},\dots,\lambda_{ij,N}),\ \gamma_{ij}=(\gamma_{ij,1},\dots,\gamma_{ij,N})'=P_{ij}^{-1}\rho_{ij},$$
$$\rho_{ij}=2(A_i')^{-1}G_j^{-1}\text{vec}(m_i-m_j),$$

while

$$k_{ij} = \frac{1}{2}\Big[\text{vec}'(m_i-m_j)G_j^{-1}\text{vec}(m_i-m_j)+\ln(|G_j|/|G_i|)\Big],$$
$$i,j=1,\dots,k,\quad j\neq i.$$

The distribution of $v_{ij}(X)$ is the same as the distribution of

$$v_{ij}(Z_{ij}) = -\frac{1}{2}\Big[Z_{ij}'(I-\Lambda_{ij})Z_{ij} - \gamma_{ij}'Z_{ij}\Big] + k_{ij} = \tag{6}$$
$$= \sum_{n=1}^{N}(b_{ij,n}Z_{ij,n}^2 + \frac{1}{2}\gamma_{ij,n}Z_{ij,n}) + k_{ij},$$

where

$$Z_{ij} = (Z_{ij,1},\dots,Z_{ij,N})' = P_{ij}'A_i\text{vec}(X-m_i),$$
$$b_{ij,n} = \frac{1}{2}(\lambda_{ij,n}-1),\quad i,j=1,2,\dots,k,\ j\neq i.$$

We have

$$v_{ij}(Z_{ij}) = \sum_{n=1}^{N_o} b_{ij,n}((Z_{ij,n}+\delta_{ij,n})^2 - \delta_{ij,n}^2)+$$
$$+\frac{1}{2}\sum_{n=N_o+1}^{N}\gamma_{ij,n}Z_{ij,n} + k_{ij},$$

where $N_o=pT_o$, $\delta_{ij,n}=\gamma_{ij,n}/4b_{ij,n}$, $i,j=1,\dots,k$, $j\neq i$.

Under condition that $X$ comes from population $\pi_i$, $\text{vec}(X-m_i)$ has a normal distribution with zero mean and covariance matrix $G_i$ and $Z_{ij}$ has a normal distribution with zero mean and the identity covariance matrix.

The expected value and the variance of the random variable $v_{ij}(Z_{ij})$ are respectively

$$E\, v_{ij}(Z_{ij}) = \sum_{n=1}^{N_o} b_{ij,n} + k_{ij} = \tfrac{1}{2}\mathrm{tr}(G_j^{-1}G_i - I) + k_{ij}, \qquad (7)$$

$$\mathrm{Var}\, v_{ij}(Z_{ij}) = \sum_{n=1}^{N_o} 2b_{ij,n}^2(1+2\varepsilon_{ij,n}^2) + \frac{1}{4}\sum_{n=N_o+1}^{N} \lambda_{ij,n}^2 = \qquad (8)$$

$$= \tfrac{1}{2}\|B\lambda_{ij}\|_N^2 - \tfrac{1}{2}\|B\|_N^2 - \mathrm{tr}(G_j^{-1}G_i - I),$$

where

$$B = \left[\begin{array}{c|c} I_{N_o} & 0 \\ \hline 0 & 0 \end{array}\right], \qquad i,j=1,2,\ldots,k,\; j\neq i.$$

Using Lapunov's central limit theorem for series (see e.g. Billingsley, 1968, Th. 7.3) one may show the following result.

Let $\lambda_{i,t}(N)$ and $\lambda_{j,t}(N)$ be the $t^{th}$ eigenvalues of of $G_i$ and $G_j$, respectively.

<u>Theorem</u>. If

a) $\sum_{t=0}^{\infty} \|\Gamma_i(t)\| = s_i < \infty$,

b) there exist $\alpha_i > 0$ such that for each $N$

$\lambda_{i,t}(N) \geq \alpha_i$, $t = 1,2,\ldots,N$,

c) there exists $\alpha_o > 0$ such that for each $N$

$|\lambda_{ij,t}(N) - 1| \geq \alpha_o$, $t=1,2,\ldots,N$,

then for all pairs $(i,j)$, $i\neq j$, as $T \to \infty$,

$$\frac{v_{ij}(X) - \tfrac{1}{2}\mathrm{tr}(G_j^{-1}G_i - I) - k_{ij}}{\left(\tfrac{1}{2}\|B\lambda_{ij}\|^2 - \tfrac{1}{2}\|B\|^2 - \mathrm{tr}(G_j^{-1}G_i - I)\right)^{1/2}} \xrightarrow{D} N(0,1).$$

Proof. Let us observe that

$$v_{ij}(Z_{ij}) - E\{v_{ij}(Z_{ij})\} = \sum_{n=1}^{N} \beta_{ij,n}(N) = S_N,$$

where

$$\beta_{ij,n}(N) = \begin{cases} b_{ij,n}[(Z_{ij,n}+\sigma_{ij,n})^2-1-\sigma^2_{ij,n}], & n=1,2,\dots,N_0, \\ \frac{1}{2}\gamma_{ij,n}Z_{ij,n}, & n=N_0+1,\dots,N, \end{cases}$$

and

$$Z_{ij}=(Z_{ij,1},\dots,Z_{ij,N})' \sim N(0,I_N).$$

Because random variables $\beta_{ij,n}(N)$ are Borel functions of independent random variables $Z_{ij,n}$, we obtain that $\beta_{ij,n}(N)$ are independent either.

Additionally

$$E\{\beta_{ij,n}(N)\} = 0, \text{ Var}\{\beta_{ij,n}(N)\}<\infty, \ n=1,\dots,N \text{ and Var}\{S_N\}>0.$$

We will proof that for $\delta=2$

$$(\text{Var } S_N)^{-2} \sum_{n=1}^{N} E\{\beta^4_{ij,n}(N)\} \xrightarrow[N\to\infty]{} 0 \quad \text{(Lapunov's condition).}$$

We have

$$E\{\beta^4_{ij,n}(N)\} = \begin{cases} b^4_{ij,n} E\{(Z_{ij,n}+\sigma_{ij,n})^2-1-\sigma^2_{ij,n}\}^4, & n=1,\dots,N_0, \\ \frac{1}{16}\gamma^4_{ij,n} E\{Z^4_{ij,n}\} & , n=N_0+1,\dots,N. \end{cases}$$

$$= \begin{cases} c_0 b^4_{ij,n} w_8(\sigma_{ij,n}) & , n=1,2,\dots,N_0, \\ \frac{3}{16}\gamma^4_{ij,n} & , n=N_0+1,\dots,N, \end{cases}$$

where $c_0$ denotessome bounding value for the central moments $^0\mu_k$, $k=0,\dots,8$ of the normal distribution $N(0,1)$ and $w_0(\cdot)$ denotesa polynomial of degree 8.

From Gershgorin's theorem (see e.g. Todd, 1978) and from condition (a) there existsthe constant $s_{0i}$ such that

$$\sup_{n,N} |\lambda_{i,n}(N)| \leq S_{0i} .$$

Moreover, from condition (b) we have

$$\sup_{n,N} |\lambda^{-1}_{j,n}(N)| \leq \alpha_j^{-1}. \tag{9}$$

Hence

$$\sup_{n,N} |\lambda_{ij,n}(N)| \leq s_{0i}\alpha_j^{-1}, \quad i,j=1,2,\ldots,k, \; j\neq i.$$

Because the expectations $\operatorname{vec} m_i$ and $\operatorname{vec} m_j$ are bounded we have that

$$\max_n |m_{i,n}-m_{j,n}| \leq c_1 ,$$

where $c_1$ denotes some constant.

Hence from (9) and condition (a) we have

$$|\gamma_{ij,n}(N)| = |\{P_{ij}^{-1}\xi_{ij}\}_n| = |\{2P_{ij}^{-1}(A_i')^{-1}G_j^{-1}\operatorname{vec}(m_i-m_j)\}_n| =$$

$$= |\{2\Lambda_{ij}P_{ij}^{-1}A_i\operatorname{vec}(m_i-m_j)\}_n| \tag{10}$$

$$\leq |\{2\Lambda_{ij}P_{ij}^{-1}A_i\operatorname{vec}(c_1 I)\}_n| \leq c_2 .$$

Using condition (c) and (10) we obtain

$$|\delta_{ij,n}(N)| = \left|\frac{\gamma_{ij,n}(N)}{4b_{ij,n}(N)}\right| = \left|\frac{\gamma_{ij,n}(N)}{2(\lambda_{ij,n}(N)-1)}\right|$$

$$\leq \left|\frac{\gamma_{ij,n}(N)}{2\inf_{n,N}(\lambda_{ij,n}(N)-1)}\right| \leq \frac{c_2}{2\alpha_0} .$$

From the above evaluation we have

$E\{\beta_{ij,n}^4(N)\} \leq c_4$, where the constant $c_4$ is independent of $n$ and $N$.

Moreover

$$\operatorname{Var} S_N = \sum_{n=1}^{N_0} 2b_{ij,n}^2(1+\delta_{ij,n}^2) + \frac{1}{4}\sum_{n=N_0+1}^{N}\gamma_{ij,n}^2 \geq \sum_{n=1}^{N_0} b_{ij,n}^2 .$$

Because $\inf_{n,N} b_{ij,n}^2(N) = \alpha_0$, then $\operatorname{Var} S_N \geq N_0\alpha_0$ and

$$(\operatorname{Var} S_N)^{-2}\sum_{n=1}^{N} E\{\beta_{ij,n}^4(N)\} \leq \frac{Nc_4}{N_0^2\alpha_0^2} \xrightarrow[N\to\infty]{} 0.$$

## REFERENCES

ANDERSON, T.W. (1958). An Introduction to Multivariate Statistical Analysis. New York. Wiley.

BILLINGSLEY, P. (1968). Convergence of Probability Measures. New York. Wiley.

KRZYŚKO, M. (1985). 'Distribution of the discriminant function.' In Linear Statistical Inference. Proceedings of the International Conference held at Poznań, Poland. June 4-8, 1984 (T. Caliński and W. Klonecki, eds), pp. 170-183. Berlin. Springer-Verlag.

SHUMWAY, R.H. (1982). 'Discriminant analysis for time series'. In Handbook of Statistics, Vol. 2 (P.R. Krishnaiah and L.N. Kanal, eds), pp. 1-46. Amsterdam. North-Holland.

TODD, J. (1978). Basic Numerical Mathematics, Vol.2 - Numerical Algebra. New York. Academic Press.

# ON EXPONENTIAL AUTOREGRESSIVE TIME SERIES MODELS

Jovan D. Mališić
University of Belgrade
Faculty of Sciences, Institute of Mathematics
Belgrade, Studentski trg 16
Yugoslavia

ABSTRACT. Some new time series models (so-called AREX(n) models) for exponential variables having $n$-th - order autoregressive structure are presented. They are analogs of the standard AR(n) model and of the EAR(n), NEAR(n) and TEAR(n) models, introduced a few years ago by Lawrance, Lewis, Gaver and the others. Some of their models can be obtained from AREX models as special cases.

The distribution of the innovation sequence (a probability mixture) and autoregressive structure of AREX processes are discussed as well.

KEYWORDS: positive-valued time series, autoregressive model, stationary, exponential marginal distribution, AREX model, probability mixture.

## 1. INTRODUCTION

It is well known that the standard linear first-order autoregressive model (AR(1) model) for a stationary sequence of random variables $\{X_n,\ n \in D = \{0,\ \pm 1,\ \pm 2,\ \dots\}\}$ is defined by the difference equation

$$X_n = \xi_n + \beta X_{n-1},\ n \in D, \qquad (1.1)$$

where $\beta$ is a parameter $(|\beta| \leq 1)$ and $\{\xi_n,\ n \in D\}$ is a sequence of independent and identically distributed (i.i.d.) random variables.

In the Gaussian AR(1) model, the innovation sequence $\{\xi_n,\ n \in D\}$ is a sequence of Gaussian random variables. But many naturally occurring time series are clearly non - Gaussian and for this reason in recent years some different models for generation of non - Gaussian time series have been constructed. For example, in many positive - valued time series we have an exponential marginal distribution. It gives a good motivation for introducing some new time series models in exponential variables.

Some earlier works on exponential time series models can be summarized as follows.



*P. Bauer et al. (eds.), Mathematical Statistics and Probability Theory, Vol. B, 147–153.*

Lawrance and Lewis (1977) discussed the stationary sequence of random variables $\{X_n, n \in D\}$ which are formed from an i.i.d. exponential sequence $\{\xi_n, n \in D\}$ according to the linear model

$$X_n = \begin{cases} \beta\,\xi_n & w.p. \quad \beta, \\ \beta\,\xi_n + \xi_{n+1} & w.p. \quad (1-\beta) \end{cases} \tag{1.2}$$

where $0 \leqq \beta \leqq 1$ and "w.p." means "with probability". This is an EMA(1) model. Gaver and Lewis (1980) in their paper [1] considered an AR model, with exponential $\mathcal{E}(\lambda)$ marginal distribution, in the form

$$X_n = \begin{cases} \rho\,X_{n-1} & w.p. \quad \rho, \\ \rho\,X_{n-1} + \xi_n & w.p. (1-\rho) \end{cases} \tag{1.3}$$

where $\rho$ is a parameter $(0 \leqq \rho \leqq 1)$ and $\xi_n$, $n \geqq 0$ are i.i.d. exponential random variables $\mathcal{E}(\lambda)$ with parameter $\lambda > 0$. This is an EAR(1) model. Jacobs and Lewis linked these two models into an EARMA(1, 1) model.

Lawrance and Lewis (1981) discussed some new AR models in exponential variables - TEAR(1) and NEAR(1):

$$X_n = \begin{cases} (1-\rho)\,\xi_n & w.p. \quad 1-\rho, \\ (1-\rho)\,\xi_n + X_n & w.p. \quad \rho \end{cases} \tag{1.4}$$

and

$$X_n = \begin{cases} \xi_n + \beta X_{n-1} & w.p. \quad \rho, \\ \xi_n & w.p. \quad 1-\rho. \end{cases} \tag{1.5}$$

Finally, Lawrance and Lewis (1985) considered the NEAR(2) model (and some its applications) as a direct generalization of the NEAR(1) model:

$$X_n = \begin{cases} \beta_1 X_{n-1} + \xi_n & w.p. \quad \alpha_1, \\ \beta_2 X_{n-2} + \xi_n & w.p. \quad \alpha_2, \\ \xi_n & w.p. \quad \alpha_3 = 1-\alpha_1 - \alpha_2. \end{cases} \tag{1.6}$$

In this paper we present a new form of time series models where marginal distributions are some exponential distributions. These forms of models (so-called AREX models) are quite distinct from earlier forms of time series models in exponential variables.

## 2. AREX(1) MODELS

Let the stationary sequence of random variables $\{X_n, n \in D\}$ be defined by the equation

$$X_n = \begin{cases} \xi_n & w.p. \quad p_o, \\ \alpha X_{n-1} & w.p. \quad p_1, \\ \beta X_{n-1}+\xi_n & w.p. \quad q_1, \end{cases} \tag{2.1}$$

where $0 \leqq p_o, p_1, q_1 \leqq 1$, $q_1 > 0$, $p_o + p_1 + q_1 = 1$, $0 < \alpha, \beta < 1$ and $\xi_n$ are some i.i.d. $\mathcal{E}(\lambda)$ random variables. Also, we suppose that $\{X_n\}$ and $\{\xi_n\}$ are "semi - independent", i.e. that $X_n$ and $\xi_m$ are independent if $n < m$.

Our first purpose is to obtain the distribution of the i.i.d. sequence $\{\xi_n\}$ which will ensure that the sequence $\{X_n\}$ in (2.1) has the exponential marginal distribution $\mathcal{E}(\lambda)$.

Let the Laplace - Stieltjes transforms of the $X$ and $\xi$ variables are denoted by

$$\phi_X(s) = E(e^{-sX}), \quad \phi_\xi(s) = E(e^{-s\xi}). \tag{2.2}$$

We know that

$$\phi_X(s) = \lambda/(\lambda + s) \tag{2.3}$$

if $X$ has $\mathcal{E}(\lambda)$ distribution. If we assume stationarity and "semi-independence", then (2.1) - (2.3) give

$$\begin{aligned} \phi_\xi(s) &= [\phi_X(s) - p_1\phi_X(\alpha s)]/[p_o + q_1\phi_X(\beta s)] = \\ &= \frac{\lambda(\lambda+\beta s)[(p_o+q_1)\lambda+(\alpha-p_1)s]}{(\lambda+s)(\lambda+\alpha s)[(p_o+q_1)\lambda+p_o\beta s]}, \end{aligned} \tag{2.4}$$

where $\alpha \geqq p_1$.

Let $\alpha$, $\beta$, $p_o$, $p_1$ and $q_1$ be choosen so that in (2.4) there are no common fractions. In this case we shall have

$$\phi_\xi(s) = (B_o\lambda)/(\lambda+s)+(B_1\lambda)/(\lambda+\alpha s)+ \frac{B_2\lambda}{(p_o+q_1)\lambda+p_o\beta s}. \tag{2.5}$$

Some computing gives

$$\begin{aligned} B_o &= (1-\beta)/(p_o+q_1-p_o\beta), \quad B_1 = p_1(\beta-\alpha)/[\alpha(p_o+q_1)-p_o\beta], \\ B_2 &= \beta q_1(\alpha-p_1-p_o\beta)(p_o+q_1)/(p_o+q_1-p_o\beta)[\alpha(p_o+q_1)-p_o\beta]. \end{aligned} \tag{2.5'}$$

It is easy to see that $B_o$, $B_1$ and $B_2/(p_o+q_1)$ are probabilities with sum equal to 1 if and only if

$$\begin{aligned} &\beta \geqq p_1/(p_1+q_1), \quad \alpha \leqq \beta \leqq \alpha/(p_o+p_1), \quad (\alpha-p_1)/p_o \leqq \beta \leqq \\ &\leqq \alpha(1 - p_1)/p_o. \end{aligned} \tag{2.6}$$

Then $\xi_n$ is the following mixture

$$\xi_n = \begin{cases} \eta_n & w.p. \quad B_o, \\ \alpha\eta_n & w.p. \quad B_1, \\ [p_o\beta/(p_o+q_1)]\eta_n & w.p. \quad B_2/(p_o+q_1) \end{cases} \tag{2.7}$$

with $\eta_n : \mathcal{E}(\lambda)$, $\lambda > 0$.

Taking the expectation of both sides in (2.1) we have

$$E(X_n) = p_oE(\xi_n)+p_1\alpha E(X_{n-1}) + q_1[\beta E(X_{n-1}) + E(\xi_n)]$$

or

$$E(\xi_n) = [1 - (p_1\alpha + q_1\beta)]/(p_o + q_1)\lambda. \tag{2.8}$$

It follows that the autocovariance function for AREX(1) time series model (2.1) is given by

$$K_r = E(X_n X_{n-r}) - EX_n \cdot EX_{n-r} = (p_1\alpha + q_1\beta)K_{r-1} \tag{2.9}$$

so that the autocorrelation function of $\{X_n\}$ is

$$\rho_r = (p_1\alpha + q_1\beta)^r, \; \rho_{-r} = \rho_r, \; \rho_o = 1. \tag{2.10}$$

## 3. SOME SPECIAL CASES

(a) <u>FAREX(1) model</u>. Let $p_o = 0$ and $0 < p_1 < 1$. Then

$$X_n = \begin{cases} \alpha X_{n-1} & w.p. \quad p_1, \\ \beta X_{n-1}+\xi_n & w.p. \quad q_1 = 1-p_1, \end{cases}$$

$$\phi_\xi(s) = \frac{(\lambda+\beta s)[q_1\lambda+(\alpha-p_1)s]}{q_1(\lambda+s)(\lambda+\alpha s)}, \quad \alpha \geqq p_1.$$

If $\alpha = p_1$ we shall have

$$\phi_\xi(\lambda) = \frac{1-\beta}{q_1}\cdot\frac{\lambda}{\lambda+s} + \frac{\beta-\alpha}{q_1}\cdot\frac{\lambda}{\lambda+\alpha s},$$

$$\xi_n = \begin{cases} \eta_n & w.p. \quad (1-\beta)/q_1, \\ \alpha\eta_n & w.p. \quad (\beta-\alpha)/q_1, \end{cases}$$

where $\eta_n : \mathcal{E}(\lambda)$, $\lambda > 0$ and $\beta \geqq \alpha$.

If $p_1 < \alpha \leqq \beta$, then

$$\phi_\xi(s) = \frac{(\alpha-p_1)\beta}{\alpha q_1} + \frac{1-\beta}{q_1}\cdot\frac{\lambda}{\lambda+\alpha} + \frac{(\beta-\alpha)p_1}{\alpha q_1}\cdot\frac{\lambda}{\lambda+\alpha s},$$

$$\xi_n = \begin{cases} 0 & w.p. \ (\alpha-p_1)\beta/\alpha q_1, \\ \eta_n & w.p. \ (1-\beta)/q_1, \\ \alpha\eta_n & w.p. \ (\beta-\alpha)p_1/\alpha q_1, \end{cases}$$

where $\eta_n : (\lambda)$, $\lambda > 0$.

Let us denote that the special case when $p_o = 0$, $\beta = \alpha = p_1$ is EAR(1) model (Lawrance and Lewis (1980)).

(b) If $\alpha = p_1 = 0$ we shall have the NEAR(1) model (Lawrance (1980)).

(c) SAREX(1) model. Let $\beta = \alpha$. Then

$$\phi_\xi(s) = \frac{\lambda[(p_o+q_1)\lambda + (\beta-p_1)s]}{(\lambda+s)[(p_o+q_1)\lambda+p_o\beta s]},$$

$$\xi_n = \begin{cases} \eta_n & w.p. \ A_1 = (1-\beta)/(p_o+q_1-p_o\beta), \\ p_o\beta\eta_n/(p_o+q_1) & w.p. \ A_2 = 1 - A_1, \end{cases}$$

where $\eta_n : \mathcal{E}(\lambda)$, $\lambda > 0$ and $\beta \geqq p_1/(p_1+q_1)$.

(d) In the case $\alpha - p_1 = p_o\beta$ we shall have

$$\phi_\xi(s) = C_1\cdot\frac{\lambda}{\lambda+s} + C_2\cdot\frac{\lambda}{\lambda+\alpha s},$$

where $C_1 = (1-\beta)/(1-\alpha)$, $C_2 = (\beta-\alpha)/(1-\alpha)$ and $\beta \geqq \alpha$.

(e) If $\alpha = \beta = p_1/(p_1+q_1)$, then $\xi_n : \mathcal{E}(\lambda)$, $\lambda > 0$.

## 4. SOME GENERALIZATIONS AND APPLICATIONS

We have shown in this paper that some well-known forms of exponential AR(1) model can be generalized. This can be done also in the following way:

$$X_n = \begin{cases} \alpha\xi_n & w.p. \ p_o, \\ X_{n-1} & w.p. \ p_1, \\ X_{n-1}+\beta\xi_n & w.p. \ q_1 \end{cases} \tag{4.1}$$

where $0 \leqq p_o, p_1, q_1 \leqq 1$, $p_o + p_1 + q_1 = 1$, $0 \leqq \alpha, \beta \leqq 1$ and $\{\xi_n, n \in D\}$ is an i.i.d. sequence of $\mathcal{E}(\lambda)$ random variables.

These generalizations can be transferred also on AREX(n), $n > 1$

models. For example, we can define SAREX(2) model in the following way:

$$X_n = \begin{cases} \xi_n & w.p.\ p_o, \\ \alpha X_{n-1} & w.p.\ p_1 \\ \alpha X_{n-1} + \xi_n & w.p.\ q_1, \\ \beta X_{n-2} & w.p.\ p_2, \\ \beta X_{n-2} + \xi_n & w.p.\ q_2, \end{cases} \tag{4.2}$$

or general AREX(2) model as:

$$X_n = \begin{cases} \alpha\xi_n & w.p.\ p_o, \\ X_{n-1} & w.p.\ p_1, \\ X_{n-1} + \beta\xi_n & w.p.\ q_1, \\ X_{n-2} & w.p.\ p_2, \\ X_{n-2} + \gamma\xi_n & w.p.\ q_2. \end{cases} \tag{4.3}$$

Finally let us note that SAREX(2) model can be written in the form

$$X_n = U_n \beta\, X_{n-1} + V_n \tag{4.4}$$

where

$$(U_n, V_n) = \begin{cases} (0, 1) & w.p.\ p_o, \\ (1, 0) & w.p.\ p_1, \\ (1, 1) & w.p.\ q_1. \end{cases}$$

So it is strictly a random coefficient autoregression. This approach gives a variety of possibilities for further developments of methods which were considered in papers on random coefficient autoregression. Also, it gives some new possibilities of model applications introduced here.

In other cases of models the situation is quite analogous.

## REFERENCES

1. D.G. Gaver and P.A.W. Lewis (1980): "First - order autoregressive Gamma sequences and point processes", *Adv. Appl. Prob.* **12**, 727-745.
2. P.A. Jacobs and P.A.W. Lewis (1977): "A mixed autoregressive - moving average exponential sequence and point process (EARMA 1,1), *Adv. Appl. Prob.* **9**, 87-104.
3. P.A. Jacobs and P.A.W. Lewis (1978): "Discrete time series generated

by mixtures. I: Correlational and runs properties", *J.R. Statist. Soc. B* **40**, 94 - 105.
4. M. Kanter (1975): "Autoregression for discrete processes mod 2", *J. Appl. Prob.* **12**, 371 - 375.
5. A.J. Lawrance (1980): "The mixed exponential solution to the first - - order autoregressive model", *J. Appl. Prob.* **17**, 546 - 552.
6. A.J. Lawrance and P.A.W. Lewis (1977): "An exponential moving average sequence and point process (EMA1)", *J. Appl. Prob.* **14**, 98-113.
7. A.J. Lawrance and P.A.W. Lewis (1980): "The exponential autoregressive - moving average EARMA ($p$, $q$) process", *J. R. Statist. Soc. B* **42**, 150 - 161.
8. A.J. Lawrance and P.A.W. Lewis (1981): "A new autoregressive time series model in exponential variables (NEAR(1))", *Adv. Appl. Prob.* **13**, 826 - 945.
9. A.J. Lawrance and P.A.W. Lewis (1985): "Modelling and residual analysis of nonlinear autoregressive time series in exponential variables", *J.R. Statist. Soc. B* **47**, 165 - 202.

# SOME CONNECTIONS BETWEEN STATISTICS AND CONTROL THEORY

Petr Mandl
Faculty of Mathematics and Physics
Charles University
Sokolovská 83
186 00 Prague 8
Czechoslovakia

ABSTRACT. Linear controlled systems with a quadratic cost function are dealt with. Using concepts of estimation theory the efficiency of a control is introduced. For systems with unknown parameters estimated by the least squares method the asymptotic distribution of the cost is derived.

The theory of controlled systems with unknown parameters has been developed by combining the methods of control theory and of mathematical statistics. One of the principles employed in the works on this subject is the method of inserting the parameter estimates into the infinite horizon optimal control (Kurano (1972), Mandl (1972)). This method was given the name Principle of Estimation and Control (see Schäl (1984)). The papers dealing with the principle are covered extensively by the existing surveys of the adaptive control theory (Kumar (1985), Pasik-Duncan (1985)).

The purpose of the present paper is twofold 1. To report about the aspects of the estimation and control method not covered by the mentioned surveys. 2. To present new results concerning the application of the method without assuming the infinite horizon optimality of the controls.

## 1. OPTIMAL CONTROL OF LINEAR SYSTEMS

We shall concentrate on linear systems with a quadratic criterion. The equation for the trajectory reads

$$(1) \qquad dX_t = fX_t dt + gU_t dt + dW_t, \qquad t \geq 0, \quad X_o = x .$$

Let the dimension of $X$ be $n$ and let $W = \{W_t, t \geq 0\}$ be the n-dimensional Wiener process with unit incremental variance matrix. The problem is to determine the control signal



*P. Bauer et al. (eds.), Mathematical Statistics and Probability Theory, Vol. B, 155–168.*

$U = \{U_t, t \in [0,T]\}$ so that the expected cost

(2) $$EC_T = E(X_T' q X_T + \int_0^T (X_t' c X_t + |U_t|^2)\, dt)$$

is minimal. The solution is

$$U_t = -g' w(t) X_t, \qquad t \in [0,T],$$

where $w(t)$ satisfies the matrix Riccati equation

(3) $$\frac{d}{dt} w + wf + f'w - wgg'w + c = 0, \qquad w(T) = q.$$

Prime denotes the transposition.

We shall consider the autonomous case. Assume the pairs of matrices $(f,g)$ and $(f', \sqrt{c})$ stabilizable. The stationary form of (3)

$$wf + f'w - wgg'w + c = 0$$

has then a unique nonnegatively definite solution. The control

(4) $$U_t = kX_t, \quad t \geqq 0, \qquad k = -g'w ,$$

yields

(5) $$E \frac{1}{T} C_T = \theta + \mathcal{O}(T^{-1}), \qquad T \to \infty ,$$

where

$$\theta = \text{trace} \quad w$$

is the minimal value achievable.

Passing beyond the text book level we realize the need to develop the control theory for systems with unknown parameters. This is an obvious interrelation to statistics.

In the case when the description of the system is not completely known to the controller, the parameter estimation is to be combined with the methods of control theory. This leads to the controls which are not of the feedback type. There fore we have to work with notions applicable rather easily to general nonanticipative controls.

A feature of advanced statistical theory is the extensive use of asymptotic methods. Consider (5) and think of

$$\bar{C}_T = \frac{1}{T} C_T$$

as of an estimate of $\theta$. A statistician would replace (5) by the consistency requirement

(6) $$\underset{T\to\infty}{\text{p lim}}\ C_T = \theta.$$

(6) is to be regarded as a property of good controls because it is the best possible result. In fact the following holds.

Proposition 1. For any $U$ such that

(7) $$\lim_{T\to\infty} \frac{1}{T} E\,|X_T|^2 = 0$$

is

$$\lim_{T\to\infty} P(\bar{C}_T \leqq \theta - \varepsilon) = 0\ , \qquad \varepsilon > 0.$$

(7) can be weakened by considering the expectation up to an event of arbitrarily small probability.

Consider next the question how to interpret the statement that $\bar{C}_T$ is an asymptotically efficient estimate of $\theta$. Asymptotic normality is to be mentioned first.

Under the feedback control (4) $\bar{C}_T$ is asymptotically normal $N(\theta, \Delta/T)$ where $\Delta$ is obtained by solving

$$x(f + gk) + (f + gk)'x + 4w^2 = 0\ ,$$

and setting

$$\Delta = \text{trace}\ x\ .$$

Asymptotic normality $N(\theta, \Delta/T)$ for $\bar{C}_T$ as $T\to\infty$ is again a property of controls which cannot be improved in the sense introduced in the next proposition.

Proposition 2. Let $U$ be such that

(8) $$\lim_{T\to\infty} \frac{1}{\sqrt{T}} E\,|X_T|^2 = 0\ .$$

Then

(9) $$\limsup_{T\to\infty} P((\bar{C}_T - \theta)\sqrt{T} \leqq y) = \Phi(y/\sqrt{\Delta}), \quad y \in (-\infty, \infty).$$

If in addition to (8)

(10) $$\underset{T\to\infty}{\text{p lim}} \frac{1}{\sqrt{T}} \int_0^T |U_t - kX_t|^2\, dt = 0\ ,$$

then

$$\text{(11)} \quad \lim_{T\to\infty} P((\bar{C}_T - \theta)\sqrt{T} \leq y) = \Phi(y/\sqrt{\Delta}), \qquad y \in (-\infty, \infty).$$

$\Phi$ denotes the distribution function of the $N(0,1)$ distribution. (9) says that asymptotically $(\bar{C}_T - \theta)\sqrt{T}$ is greater or equal to a random variable with the $N(0,\Delta)$ distribution in the sense of stochastic ordering. We can call the controls under which (11) holds <u>asymptotically efficient</u>.

The approaches from mathematical statistics lead us to the discovery that also other limit theorems define an optimality property. Let us mention here the arcsine law. The quantity

$$B_T = \frac{1}{T}\int_0^T \chi\{\bar{C}_t > \theta\}\, dt$$

is the average time spent by $\bar{C}_t$ above $\theta$. $B_T$ is small for good controls. The attainable bound is given by the arcsine distribution.

<u>Proposition 3</u>. Let (8) hold. Then

$$\limsup_{T\to\infty} P(B_T \leq y) \leq \frac{2}{\pi} \arcsin\sqrt{y}, \quad y \in [0,1].$$

If (8), (10) holds, then

$$\text{(12)} \quad \lim_{T\to\infty} P(B_T \leq y) = \frac{2}{\pi} \arcsin\sqrt{y}, \qquad y \in [0,1].$$

Example 1

Let us apply the introduced concepts to self-optimizing controls of systems with unknown parameters. For the sake of clarity consider here the simpliest case that $f$ in (1) depends on a one-dimensional parameter $\alpha$,

$$f(\alpha) = f_0 + \alpha f_1, \qquad \alpha \in R^1,$$

while the other matrices in (1), (2) are constant. We have thus a family of systems as described by (1), (2).

Denote the optimal feedback gain by $k(\alpha)$ and the optimal average cost by $\theta(\alpha)$.

Let $\alpha_0$ be the true value of the parameter, and assume that it is unknown to the controller. He can proceed using the estimation and control principle as follows. From the observation of $X_s$, $s \leq t$, he computes an estimate $\alpha_t^*$ of $\alpha_0$. The substitution of $\alpha_t^*$ for the true parameter value leads from (4) to

$$U_t = k(\alpha_t^*)\, X_t\,, \qquad t \geq 0\,. \tag{13}$$

The control (13) has the self-optimizing property if $\alpha_t^*$ is strongly consistent.

Let us take for $\alpha_t^*$ the maximum likelihood estimate of $\alpha_0$. The logarithmic likelihood function for the observation of $X_t$, $t \in [0,T]$, is

$$L_T(\alpha) = \int_0^T (f(\alpha)\, X_t + gU_t)'\, dX_t - \frac{1}{2} \int_0^T |f(\alpha) X_t + gU_t|^2 dt.$$

$L_T$ is the logarithm of the probability density with respect to the n-dimensional Wiener measure starting from x. The solution of

$$\frac{d}{d\alpha} L_T(\alpha_T^*) = 0$$

yields

$$\alpha_T^* = \left(\int_0^T X' f_1'\, dX - \int_0^T X' f_1' (f_0 X + gU)\, dt\right) / Q_T =$$

$$= \alpha_0 + \int_0^T X' f_1'\, dW / Q_T = \alpha_0 + \mathcal{W}_{Q_T} / Q_T$$

where

$$Q_T = \int_0^T |f_1 X|^2\, dt\,.$$

$\mathcal{W}$ is a Wiener process. An application of the strong law of large numbers to the last term shows that

$$\int_0^\infty |f_1 X_t|^2 \, dt = \infty \qquad \text{a.s.} \tag{14}$$

implies the strong consistency of $\alpha_T^*$, i.e.

$$\lim_{T \to \infty} \alpha_T^* = \alpha_0 \qquad \text{a.s.}$$

a.s. means almost surely.

Liapounov type conditions, which are always valid if the parameters is restricted to a neighbourhood of $\alpha_0$, can be imposed to ensure the validity of (14).

Proposition 4. Let the control (13) based on the maximum likelihood estimation guarantee (14). Then it is asymptotically efficient in the sense of (11) or (12).

It does not require much effort to see that estimates regarding the moments of $\bar{C}_T$ are difficult to establish while the convergence in distribution is a feasible tool.

## 2. QUADRATIC COST AND SELF-TUNING CONTROL

Objections were raised against the use of optimal stationary controls in combination with the parameter estimation. One of them is the computational complexity. It must also be noted that controls designed according to other rules (e.g. the pole shifting) are more frequently used by the control engineers. Let us thus investigate the quadratic cost functionals without postulating the average cost optimality of the controls.

The basic model is a general version of that considered in Example 1. Let the trajectory of the controlled system fulfil

$$dX_t = f(\alpha) X_t dt + gU_t dt + dW_t , \quad t \geq 0, \quad X_0 = x , \tag{15}$$

where

$$f(\alpha) = f_0 + \alpha^1 f_1 + \dots + \alpha^q f_q , \quad \alpha = (\alpha^1, \dots, \alpha^q) \in R^q.$$

Let $W = \{W_t, t \geq 0\}$ be the Wiener process with incremental variance matrix $h$,

$$dW_t \, dW_t' = h \, dt .$$

To each parameter $\alpha$ there corresponds a feed-back gain $k(\alpha)$ computed by applying some rule to (15). It is desired to have

$$dX_t = f(\alpha)X_t dt + gk(\alpha)X_t dt + dW_t , \qquad t \geq 0,$$

provided that $\alpha$ coincides with the true parameter value. The latter will be denoted by $\alpha_0$. It is assumed to be unknown to the controller.

From the observation of $X_t$, $t \in [0,T]$, the estimate $\alpha_T^*$ of $\alpha_0$ is obtained by the least squares method. To see what this means in the case of continuous observation take first the discretized version of (15)

$$\Delta X_{t_k} = f(\alpha) X_{t_k} \Delta t_k + gU_{t_k} \Delta t_k + \Delta W_{t_k}, \qquad k = 0,1,\ldots$$

The weighted sum of squares to be minimized is then

$$(16) \quad \sum_k \frac{1}{\Delta t_k} (\Delta X_{t_k} - f(\alpha)X_{t_k} \Delta t_k - gU_{t_k} \Delta t_k)'$$

$$l(\Delta X_{t_k} - f(\alpha)X_{t_k} \Delta t_k - gU_{t_k} \Delta t_k) .$$

$l$ is a nonnegatively definite symmetric matrix. Note that $\Delta t_k$ is proportional to the observation error. The least squares estimate $\alpha^*$ obtained from (16) fulfils the equations

$$\sum_j \sum_k X_{t_k}' f_i' l f_j X_{t_k} \Delta t_k \alpha^{*j} = \sum_k X_{t_k}' f_i' l(\Delta X_{t_k} - f_0 X_{t_k} \Delta t_k - gU_{t_k} \Delta t_k),$$

$$i = 1,\ldots q.$$

From here letting $\Delta t_k \to 0$ we get the system of equations for $\alpha_T^*$

$$(17)\quad \sum_j \int_0^T X_t' f_i' l f_j X_t dt\, \alpha_t^{*j} = \int_0^T X_t' f_i' l (dX_t - f_0 X_t dt - gU_t dt),\ i=1,\dots q,$$

and

$$(18)\quad \sum_j \int_0^T X_t' f_i' l f_j X_t dt (\alpha_T^{*j} - \alpha_0^j) = \int_0^T X_t' f_i' l\ dW_t, \quad i=1,\dots,q .$$

$\alpha_T^*$ is the maximum likelihood estimate if $l$ is the inverse of $h$. This was the case in Example 1.

$\alpha_0$ being the true value of $\alpha$, the aimed performance equation is

$$(19)\quad dX_t = (f(\alpha_0) + gk(\alpha_0))\, X_t\, dt + dW_t, \quad t \geqq 0.$$

The self-tuning regulator designed according to the estimation and control principle leads to

$$(20)\quad dX_t = (f(\alpha_0) + gk(\alpha_t^*)) X_t dt + dW_t, \quad t \geqq 0 .$$

(20) has the self-tuning property if

$$(21)\quad \lim_{t\to\infty} \alpha_t^* = \alpha_0 \qquad \text{a.s.}$$

provided that $k(\alpha)$ is continuous at $\alpha_0$. From (18) it is seen that the hypothesis (21) is not a stringent one. For $q = 1$ (14) is sufficient. (See also Duncan (1985).)

In the sequel we shall study the asymptotic behaviour of quadratic functionals

$$C_T = \int_0^T (X_t' cX_t + |U_t|^2)\, dt$$

as $T \to \infty$ without assuming the optimality property of $k(\alpha)$. The purpose is to clarify the connections between quadratic cost optimality and other rules for the design of regulators and to provide a numerical characterization of the speed of the self-tuning. This is achieved by demonstrating an invariance principle from which also some results stated in Section 1 follow.

Let us write briefly

$$f = f(\alpha_0) , \qquad k = k(\alpha_0) .$$

We assume that (19) is stable. Then the limiting distribution of $X_t$ is $N(0,v)$ where the variance matrix $v$ fulfils

$$(22) \qquad (f + gk)v + v(f + gk)' + h = 0 .$$

Set

$$\theta = \operatorname{trace} ((c + k'k)v) .$$

$\theta$ is the limiting average cost, and the cost potential for starting position $X_0 = x$ has the expression

$$\int_0^\infty E_x(X_t'(c+ k'k)X_t - \theta)dt = x'wx + \text{const.}$$

$w$ is the unique symmetric matrix such that

$$(23) \qquad w(f + gk) + (f + gk)'w + c + k'k = 0 .$$

Let

$$U_t = K_t X_t , \qquad t \geq 0 ,$$

be a nonanticipative control, and let the gain matrices $\{K_t, t \geq 0\}$ have piecewise continuous trajectories. The following equality makes from the cost potential a useful tool to study $C_T$ as $T \to \infty$ .
It holds

$$(24) \quad C_T - \theta T + X_T'wX_T - x'wx - \int_0^T X_t'(K_t+k+2g'w)'(K_t-k)X_t \, dt =$$

$$= 2 \int_0^T X_t' \, w \, dW_t , \qquad T \geq 0 .$$

To verify (24) the Itô formula gives us

$$X_T'wX_T - x'wx = \int_0^T dX'wX = 2\int_0^T X'w(f + gK) X \, dt +$$

$$+2 \int_0^T X'wdW + \operatorname{trace} (wh) T.$$

Using (23) one gets from here

$$C_T - \mathrm{trace}(wh)T + X_T'wX_T - x'wx - \int_0^T |KX|^2 dt + \int_0^T |kX|^2 dt -$$

$$- 2 \int_0^T X'wg(K-k)Xdt = 2 \int_0^T X'wdW.$$

From here for $K_t = k$, $t \geq 0$ we infere that

$$\mathrm{trace}\ (wh) = \mathrm{trace}\ ((c+k'k)v) = \theta,$$

and hence (24) follows.

Before stating the invariance principle let us introduce some denotation. Set

$$a_{ij} = \mathrm{trace}\ (f_i'lf_jv), \qquad a = \|a_{ij}\|_{i,j=1}^{q},$$

$$b_{ij} = \mathrm{trace}\ (f_i'lhlf_jv), \qquad b = \|b_{ij}\|_{i,j=1}^{q},$$

$$p^i = 2\mathrm{trace}\ (whlf_iv), \qquad p' = (p^1, p^2, \ldots, p^q),$$

$$k_i = \frac{\partial}{\partial \alpha^i} k(\alpha_0),$$

$$r^i = 2\mathrm{trace}\ ((k'+wg)k_iv), \quad r' = (r^1, r^2, \ldots, r^q),$$

$$d = 4\mathrm{trace}\ (whwv).$$

<u>Proposition 5</u>. Let the matrix a be nonsingular and let $k(\alpha)$ be twice continuously differentiable at $\alpha_0$. Assume

(25) $$U_t = k(\alpha_t^*)X_t, \quad t \geq t_0,$$

where $\alpha_t^*$ is the least squares estimate of $\alpha_0$ satisfying

(26) $$\lim_{t\to\infty} \alpha_t^* = \alpha_0 \quad \text{a. s.}$$

Then the probability distribution of

$$(\bar{C}_{sT} - \theta s)\sqrt{T}, \qquad s \in [0,1],$$

converges weakly as $T\to\infty$ to the distribution of

$$Z_s^1 + \int_0^s \frac{Z_t^2}{t}\, dt \ , \quad s \in [0,1] \ , \tag{27}$$

where $Z = \{ (Z_t^1, Z_t^2)', \ t \in [0,1] \}$ is the two-dimensional Wiener process with incremental variance

$$d(ZZ') = \begin{pmatrix} d & , p'a^{-1}r \\ p'a^{-1}r, & r'a^{-1}ba^{-1}r \end{pmatrix} dt.$$

Proof. (26) implies the validity of the strong law of large numbers for $C_T$ and analogous quadratic functionals of the trajectory. To prove this the Liapounov function methods of Mandl (1987b) should be applied to (24). Moreover

$$\lim_{T\to\infty} \frac{|X_T|^2}{\sqrt{T}} = 0 \quad \text{a. s.}$$

In particular

$$\left\| \int_0^T X_t' f_i' l f_j X_t \, dt \right\|_{i,j=1}^{q} \sim aT \ , \quad T \to \infty \ .$$

Hence, from (18),

$$\alpha_t^* - \alpha_0 \sim t^{-1} a^{-1} \left( \int_0^t X' f_1' l dW, \ldots, \int_0^t X' f_q' l dW \right)', t \to \infty. \tag{28}$$

The random processes

$${}^T Y = \{ ( {}^T Y_s^0, {}^T Y_s^1, \ldots, {}^T Y_s^q )', \ s \in [0,1] \} =$$

$$= \left\{ \left( \frac{2}{\sqrt{T}} \int_0^{sT} X' w dW, \frac{1}{\sqrt{T}} \int_0^{sT} X' f_1 l dW, \ldots, \frac{1}{\sqrt{T}} \int_0^{sT} X' f_q l dW \right)', \ s \in [0,1] \right\}$$

converge weakly as $T \to \infty$ to the Wiener process $\mathcal{W} = \{\mathcal{W}_s, s \in [0,1]\}$ with incremental variance matrix

$$\begin{pmatrix} d & , p' \\ p & , b \end{pmatrix} \ .$$

The convergence is established by taking arbitrary linear combination of the stochastic integrals and expressing it by means of a random time change in a Wiener process ( see Mc. Kean (1969)). The time scale is given by the quadratic variation to which the strong law of large numbers applies.

In (24) we have

$$X_t'(K_t+k+2g'w)'(K_t-k)X_t = \sum_i 2X_t'(k'+wg)k_iX_t(\alpha_t^{*i}-\alpha_0^i) +$$
$$+ \mathcal{O}(|X_t|^2 \cdot |\alpha_t^* - \alpha_0|^2) \quad , \quad t \to \infty \quad .$$

Set

$$u_i = 2(k' + wg)k_i \quad .$$

From (24) follows

$$(29) \quad C_T - \theta T = 2\int_0^T X_t' w dW_t + \sum_i \int_0^T X_t' u_i X_t(\alpha_t^{*i} - \alpha_0^i)dt + o_p(\sqrt{T}).$$

Let

$$Z_t^i = t(\alpha_t^{*i} - \alpha_0^i)$$

and recall that

$$(30) \qquad \lim_{t\to\infty} \frac{1}{t}\int_0^t X_s' u_i X_s ds = r_i \quad \text{a. s.}$$

Integration by parts yields

$$\int_0^T X_t' u_i X_t(\alpha_t^{*i}-\alpha_0^i)dt = \int_0^T X_t' u_i X_t \frac{Z_t^i}{t}dt = \frac{1}{T}\int_0^T X_t' u_i X_t dt Z_T^i -$$
$$- \int_0^T \frac{1}{t}\int_0^t X_s' u_i X_s ds dZ_t^i + \int_0^T \frac{1}{t^2}\int_0^t X_s' u_i X_s ds Z_t^i dt =$$
$$= \int_0^T \frac{1}{t^2}\int_0^t X_s' u_i X_s ds Z_t^i dt + o_p(\sqrt{T}) \quad , \quad T \to \infty \; .$$

With regard to (28) we obtain from (29)

$$(\bar{C}_{sT} - \theta s)\sqrt{T} = {}^T Y_s^0 + \sum_{i=1}^q \int_0^s \frac{1}{zT}\int_0^{zT} X_z' u_i X_z dz \frac{1}{t} e_i a^{-1}\, {}^T Y_t^i dt +$$
$$+ o_p(\sqrt{T}) \quad , \quad s \in [0,1] \; .$$

$e_i$ is the row vector having 1 at i-th position and 0 elsewhere. Taking into account (30) and the weak convergence of ${}^T Y$ to $\mathcal{W}$ we infere that $(\bar{C}_{sT} - \theta s)\sqrt{T}$ converges weakly to

$$\mathcal{W}_s^0 + \int_0^s \frac{1}{t}\sum_{i=1}^q r^i e_i a^{-1} \mathcal{W}_t^i dt, \quad s \in [0,1] \; ,$$

which coincides with (27). □

<u>Corollary 1</u>. Under the assumptions of Proposition 5 the asymptotic distribution of $(\bar{C}_T - \theta)\sqrt{T}$ as $T\to\infty$ is normal $N(0,\Delta)$

with

(31) $$\Delta = d + 2(p + ba^{-1}r)'a^{-1}r \ .$$

The asymptotic variance parameter has the interpretation as measure of risk associated with the cost function $C_T$. For (19) the parameter coincides with d. Thus, if $\Delta < d$ we can say that (20) is more risk averting than (19).

Remark 1. If $k(\alpha)$ are optimal stationary controls, then

$$k' + wg = 0.$$

(27) reduces to $Z_s^1$, $s \in [0,1]$, and the validity of (11) and (12) obtains from the invariance principle.

Example 2

Let the system dynamics be

(32) $$\frac{d^3}{dt^3} X_t + \alpha^2 \frac{d^2}{dt^2} X_t + \alpha^1 \frac{d}{dt} X_t + \alpha^0 X_t = \dot{W}_t , \qquad t \geq t_0 ,$$

where $\dot{W}$ is the standard white noise. It is desired to have

$$\frac{d^3}{dt^3} X_t + \beta^2 \frac{d^2}{dt^2} X_t + \beta^1 \frac{d}{dt} X_t + \beta^0 X_t = \dot{W}_t , \qquad t \geq t_0 .$$

The control signal is therefore

$$U_t = (\beta^0 - \alpha_t^{0*})X_t + (\beta^1 - \alpha_t^{1*})\frac{d}{dt} X_t + (\beta^2 - \alpha_t^{2*})\frac{d^2}{dt^2} X_t , \quad t \geq t_0 .$$

The estimate $\alpha_t^*$ obtains by maximizing the integrated square of the left-hand side of (32).

Let

$$\alpha_0^0 = 1 , \ \alpha_0^1 = 2 , \ \alpha_0^2 = 2 , \ \beta^0 = 2 , \ \beta^1 = 4 , \ \beta^2 = 3 ,$$

$$C_T = \int_0^T ( cX_t^2 + U_t^2 ) \, dt , \quad c \geq 0 .$$

$\theta$ and the parameters of (31) as function of c are

$$\theta = 0.075c + 0.375 ,$$

$$d = 0.025875c^2 + 0.09375c + 0.171875 ,$$

$$\Delta = 0.025875c^2 - 0.22625c + 2.171875 .$$

For $c > 6.25$ (20) is more risk averting than (19).

## REFERENCES

DAVIS, M.H.A. and VINTER, R.B. (1984). Stochastic Modelling and Control. Chapman and Hall, London.

DUNCAN, T.E. and PASIK-DUNCAN, B. (1985). Adaptive control of continuous time linear systems. Preprint, University of Kansas.

MC. KEAN, H.P. (1969). Stochastic Integrals. Academic Press, New York.

KUCERA, V. (1973). A review of the matrix Riccati equation. Kybernetika (Prague) 9, 42-61.

KUMAR, P.R. (1985). A survey of some results in stochastic adaptive control. SIAM J. Control and Optimization 23, 329-380.

KURANO, M. (1972). Discrete-time Markovian decision processes with an unknown parameter:average return criterion. J. Op. Research Soc. Japan 15, 67-76.

MANDL, P. (1972). An application of Itô's formula to stochastic control systems. In Lecture Notes in Mathematics 294, Springer-Verlag, Berlin, 8-13.

MANDL, P. (1986). Asymptotic ordering of probability distributions for linear controlled systems with quadratic cost. In Lecture Notes in Control and Inf. Sc. 78, Springer-Verlag, Berlin, 277-283.

MANDL, P. (1987a). Limit theorems of probability theory and optimality in linear controlled systems with quadratic cost. In Proc. of 5th IFIP Working Conference on Stochastic Differential Systems. To appear.

MANDL, P. (1987b). On transient phenomena in self-optimizing control systems. In Trans. of 10th Prague Conference on Information Theory etc. To appear.

PASIK-DUNCAN, B. (1985). On adaptive control. Preprint. Central School of Planning and Statistics, Warsaw.

SCHÄL, M. (1984). Asymptotic results for sequential Markov decision models under uncertainty. Statistics and Decisions 2, 39-62.

# MAXIMUM WAITING TIME WHEN THE SIZE OF THE ALPHABET INCREASES

Tamás F. Móri
Department of Probability Theory,
Loránd Eötvös University,
Budapest, Muzeum krt. 6-8.
H-1088 Hungary

ABSTRACT. Observing a sequence of random letters coming from a finite alphabet we consider the waiting time till each of a given subset of length k words occurs as a run. A limit theorem is derived for this waiting time as the size of the alphabet and of the given set of words tends to infinity. The main difficulty compared with the classical case k=1 is the presence of overlapping between the words.

## 1. INTRODUCTION AND RESULTS

Let $\mathfrak{X} = \mathfrak{X}_n$ be a finite alphabet of size n. Consider a sequence $X_1, X_2, \ldots$ of i.i.d. random variables distributed uniformly in $\mathfrak{X}$. We are interested in the runs ( = connected subsequences) of prescribed length observed in the course of the process $X_i$. Let $k=k_n$ be a sequence of positive integers and let $\mathfrak{X}^k$ denote the set of length k words over $\mathfrak{X}$. For any word $A \in \mathfrak{X}^k$ let

$$T(A) = \min \{ m: (X_{m-k+1} X_{m-k+2} \ldots X_m) \equiv A \} ,$$

the waiting time till A is observed as a run. Finally, given a set of words $H_n \subset \mathfrak{X}_n^k$ for every n, define

$$W(H_n) = \max \{ T(A): A \in H_n \} ,$$

the waiting time till each word of $H_n$ occurs as a run.

The aim of the present note is to find the limit distribution of $W(H_n)$ as n and $|H_n|$ tend to infinity (and k may also vary moderately with n). The case where the alphabet was fixed and the length of the words as well as the size of the given set of words tended to infinity was studied in [4] and [5]. The present case is somewhat simpler, because the overlapping between the words, which constitutes the main difficulty of the problem, decreases as the size of the



P. Bauer et al. (eds.), Mathematical Statistics and Probability Theory, Vol. B, 169–178.

alphabet increases. We get estimations for the rate of convergence, too.

THEOREM

(a) *Suppose* $|H_n| \le n/\log^4 n$. *Then with an absolute constant* C

$$\sup_y |P(n^{-k}W(H_n) - \log|H_n| \le y) - e^{-e^{-y}}| \le C|H_n|^{-1/2}\log|H_n|.$$

(b) *Suppose* $n/\log^4 n < |H_n| \le \exp(n^{1/3})$. *Then with an absolute constant* C

$$\sup_y |P(n^{-k}W(H_n) - \log|H_n| \le y) - e^{-e^{-y}}|$$

$$\le C(n|H_n|)^{-1/4}\log^2(n|H_n|).$$

## 2. REMARKS

2.1. For $|H_n| \sim n/\log^4 n$ both estimates give $O(n^{-1/2}\log^3 n)$, so the above theorem can be formulated in a unified way: at any time that bound is to be applied whose order of magnitude is greater.

2.2. In the most important special case of the considered problem k is fixed and $H_n = \mathfrak{X}_n^k$. The above theorem asserts that the limiting distribution function is $F(y) = \exp(-\exp(-y))$ and the rate of convergence is at least $O(n^{-(k+1)/4}\log^2 n)$. This bound is not sharp but further improvements for k>1 require great effort while our proof is quite simple. For k=1 we arrive at the following classical urn problem [1]. Placing balls into n urns at random, one after another, find the limit distribution as $n\to\infty$ of the number of balls needed till there is at least one ball in every urn. In that case the lack of overlapping between the one-letter words enables us to give sharp estimations:

$$\lim_{n\to\infty} \frac{n}{\log n} \sup_y |P(n^{-1}W(\mathfrak{X}_n) - \log n \le y) - e^{-e^{-y}}|$$

$$= 0.83996 \qquad (1)$$

The proof of this assertion is sketched after the proof of the Theorem.

2.3. In the complementary case, i.e. where $|H_n| > \exp(n^{1/3})$, k has to grow with n rapidly: $k>n^{1/3}(\log n)^{-1}$. This case can be treated by the method developed in [5], basing on the graph-sieve of Rényi. Though originally made for fixed size alphabet, the proof of Theorem 1, loc.cit., remains valid for varying n, too, and it yields an $O(|H_n|^{-1/6}\log^2|H_n|)$ bound.

## 3. PROOF OF THE THEOREM

For every $A\in H_n$ let $C(A)$ denote the event

$$\{T(A)>n^k(\log|H_n|+y)\}.$$

Then

$$\{n^{-k}W(H_n)-\log|H_n|\le y\} = \bigcap_{A\in H_n} \overline{C(A)},$$

where dash is for complement. Now by the inclusion-exclusion formula

$$|P(\bigcap_{A\in H_n} \overline{C(A)}) - \sum_{r=0}^{m-1}(-1)^r S_r| \le S_m , \quad m<|H_n|, \tag{2}$$

with $S_0=1$ and

$$S_r=\sum P(C(A_1)\cap \dots \cap C(A_r))=\sum P(\min_{1\le i\le r} T(A_i)>n^k(\log|H_n|+y)),$$

where the summation runs over the size r subsets of $H_n$.

The terms of $S_r$ can be estimated by the help of [3, Lemma 3]. With the notation $Z=\min\{T(A_1),\dots,T(A_r)\}$ we have

$$\exp(-(1+2rkn^{-k})x)\le P(Z>E(Z)x)\le \exp(2rkn^{-k}-x). \tag{3}$$

For the calculation of $E(Z)$ Theorem 1 of [2] provides the following quick algorithm. Given two words $A=(a_1\dots a_k)$ and $B=(b_1\dots b_k)$ define the *leading number* of A over B as $A*B=\sum_{i=1}^{k}\varepsilon_i n^i$, where $\varepsilon_i=1$ if B overlaps A in length i, that is, $(a_{k-i+1}\dots a_k)\equiv(b_1\dots b_i)$, and $\varepsilon_i=0$ otherwise. Obviously,

$$n^k\le A*A\le n^k+n^{k-1}+\dots+n<\frac{n}{n-1}n^k, \text{ and } A*B<\frac{1}{n-1}n^k \text{ for } A\ne B.$$

Now, for any r-subset of words $\{A_1, \dots, A_r\}$ let $p_i=P(Z=T(A_i))$, $1\le i\le r$, then

$$p_1A_1*A_j+p_2A_2*A_j+\dots+p_rA_r*A_j=E(Z), \quad 1\le j\le r, \tag{4}$$

which together with the relation $p_1+p_2+\ldots+p_r=1$ gives a non-singular system of linear equations for the r+1 unknowns $p_1, \ldots, p_r$, E(Z). Especially, E(T(A))=A*A.

We shall perform the proof simultaneously for both (a) and(b). For the sake of brevity let $F(y)=\exp(-\exp(-y))$ and $H=|H_n|$. First, let y be bounded by

$$K^{-1} \le F(y) \le 1-K^{-1},$$

where

$$K=H^{1/2}/\log H, \tag{5a}$$

$$K=(nH)^{1/4}/\log^2(nH), \tag{5b}$$

resp. Clearly in both cases $K^{-1}$ is uniformly bounded and

$$e^{-y}=-\log F(y) \le \log K \quad \text{and} \quad y< -\log(1-F(y)) \le \log K.$$

Further, let m=[4 log K] where brackets stand for integer part, and let $\varepsilon=m(nH)^{-1/2}$. Then

$$m< 2\log H, \qquad \log(HK)<\frac{3}{2}\log H, \tag{6a}$$

$$m< \log(nH), \qquad \log(HK)<\frac{5}{4}\log(nH). \tag{6b}$$

For the upper estimate of $S_r$ let us divide the sum into two: the first part, denoted by $S_r'$, contains all r-subsets of words $A_1, \ldots, A_r$ belonging to $H_n$, for which

$$A_i*A_i \le (1+\varepsilon)n^k, \quad A_i*A_j \le n^k\varepsilon \quad (i,j=1,2,\ldots,r,\ i\ne j),$$

the rest is contained in $S_r''$.

Now we show that for $r\le m$ the number of summands of $S_r''$ is less than

$$\frac{4}{r!}H^r m^2(\varepsilon H)^{-1}$$

$$\le \frac{8}{r!}H^r n^{1/2}K^{-1}, \tag{7a}$$

$$\le \frac{4}{r!}H^r n(K^2\log^3(nH))^{-1}, \tag{7b}$$

where (7) follows from (5), (6) and the definitions of m and ε.

First, let us estimate the number of words A for which $A*A>(1+\varepsilon)n^k$. Such an A has to overlap itself at least in length ℓ, where $n^k<n+n^2+\ldots+n^\ell<2n^\ell$, and the number of length k words that overlap themselves in length ℓ is just $n^{k-\ell}$, hence the number of words A with the desired property is bounded by $n^{k-\ell}+n^{k-\ell-1}+\ldots+n < 2n^{k-\ell} < 4/\varepsilon$. Similarly one can show that the same bound holds for the number of words $B \neq A$ such that $A*B > \varepsilon n^k$ with a fixed A. Now any r-subset in $S_r''$ contains either a word or a pair with relatively large leading number. Clearly, there are at most $\binom{H}{r-1}4/\varepsilon$ of those containing a bad word, and at most $\binom{H}{r-1}(r-1)4/\varepsilon$ of those containing a bad pair. Hence (7) follows.

Terms of $S_r'$ and $S_r''$ can be estimated in the following way. Using (4) and the inequalities immediately preceding (4) one can readily verify that for r-subsets in $S_r'$

$$\frac{1}{r}n^k \le E(Z) \le \left(\frac{1}{r}+\varepsilon\right)n^k,$$

while for r-subsets in $S_r''$

$$\frac{1}{r}n^k \le E(Z) \le \left(\frac{1}{r}+\frac{1}{n-1}\right)n^k \le \left(\frac{1}{r}+\frac{2}{n}\right)n^k$$

holds. Then inequality (3) applies to the terms of $S_r'$ :

$$\begin{aligned} &P(Z>n^k(\log H+y)) \\ &\le \exp\left(2rkn^{-k}-\left(\frac{1}{r}+\varepsilon\right)^{-1}(\log H+y)\right) \\ &\le \exp(2rkn^{-k}-r(1-r\varepsilon)(\log H+y)) \\ &= H^{-r}e^{-ry}\exp(2rkn^{-k}+r^2\varepsilon\log(HK)). \end{aligned}$$

In case (a) $2kn^{-k} \le \frac{2}{\log n}\,\frac{\log n^k}{n^k} \le 3\,\frac{\log H}{H}$ and $H\log^4 H < H\log^4 n \le n$, hence

$$\begin{aligned} &P(Z>n^k(\log H+y)) \\ &\le H^{-r}e^{-ry}\exp(6H^{-1}\log^2 H+12H^{-1}\log^2 H) \\ &\le H^{-r}e^{-ry}(1+C_1K^{-2}). \end{aligned} \tag{8a}$$

In case (b) $2kn^{-k} \le \frac{2}{\log n}\,\frac{\log\sqrt{nH}}{\sqrt{nH}} \le \frac{3}{2}(nH)^{-1/2}\log(nH)$, hence

$$\begin{aligned} &P(Z>n^k(\log H+y)) \\ &\le H^{-r}e^{-ry}\exp\left(\tfrac{3}{2}(nH)^{-1/2}\log^2(nH)+\tfrac{5}{4}(nH)^{-1/2}\log^4(nH)\right) \\ &\le H^{-r}e^{-ry}(1+C_2K^{-2}). \end{aligned} \tag{8b}$$

Terms of $S_r''$ can be estimated in a similar way:

$$P(Z>n^k(\log H+y))$$
$$\le \exp(2rkn^{-k}-(\tfrac{1}{r}+\tfrac{2}{n})^{-1}(\log H+y))$$
$$\le \exp(2rkn^{-k}-r(1-\tfrac{2r}{n})(\log H+y))$$
$$= H^{-r}e^{-ry}\exp(2rkn^{-k}+\tfrac{2}{n}r^2\log(HK)).$$

Here $2rkn^{-k}\le\frac{2r}{n}<\frac{2}{n}r^2\log(HK)$. In case (a) $\frac{4}{n}r^2\log(HK)$ $\le\frac{6}{n}\log^3 H\le\frac{6}{\sqrt{n}}K^{-1}$, hence

$$P(Z>n^k(\log H+y))\le H^{-r}e^{-ry}(1+C_3n^{-1/2}K^{-1}). \tag{9a}$$

In case (b) $\frac{4}{n}r^2\log(HK)\le\frac{5}{n}\log^3(nH)\le\frac{5}{n}(\log n+n^{1/3})^3<50$, thus

$$P(Z>n^k(\log H+y))\le H^{-r}e^{-ry}(1+C_4\tfrac{1}{n}\log^3(nH)). \tag{9b}$$

The number of terms of $S_r$ is less than $\frac{1}{r!}H^r$, hence by (6)-(9)

$$S_r\le\frac{1}{r!}e^{-ry}(1+C_5K^{-2}). \tag{10}$$

The lower estimation for $S_r$ is much simpler. The number of terms of $S_r$ exceeds

$$\frac{1}{r!}(H-r)^r\ge\frac{1}{r!}H^r(1-r^2/H),$$

while the terms themselves are bounded from below by

$$\exp(-(1+2rkn^{-k})r(\log H+y))$$
$$\ge H^{-r}e^{-ry}\exp(-2r^2kn^{-k}\log(HK))$$
$$\ge H^{-r}e^{-ry}(1-2r^2kn^{-k}\log(HK)).$$

Here we used (3) again. Hence

$$S_r\ge\frac{1}{r!}e^{-ry}(1-r^2H^{-1}-2r^2kn^{-k}\log(HK)).$$

In case (a) $r^2H^{-1}\le 4K^{-2}$ and

$$2r^2kn^{-k}\log HK\le 3r^2n^{-1}\log H\le 12(H\log H)^{-1}\le 12K^{-2}.$$

In case (b)

$$H^2\log^4(nH) \geq H^2\log^4 n \geq nH,$$

thus $H \geq (nH)^{1/2}/\log^2(nH)$, hence $r^2H^{-1} \leq K^{-2}$. Further,

$$2r^2kn^{-k}\log(HK) \leq \frac{3}{2}(nH)^{-1/2}\log^3(nH)\,\frac{5}{4}\log(nH) < 2K^{-2}.$$

In both cases

$$S_r \geq \frac{1}{r!}e^{-ry}(1-C_6K^{-2}). \tag{11}$$

Returning now to (2) we can write

$$\begin{aligned}\Delta(y) &= |P(n^{-k}W(H_n)-\log|H_n|\leq y)-F(y)| \\ &\leq \Big|P\Big(\bigcap_{A\in H_n}\overline{C(A)}\Big)-\sum_{r=0}^{m-1}(-1)^r S_r\Big| + \\ &\quad +\Big|\sum_{r=0}^{m-1}(-1)^r\, S_r-\frac{1}{r!}e^{-ry}\Big| + \\ &\quad +\Big|\sum_{r=0}^{m-1}(-1)^r\frac{1}{r!}e^{-ry}-F(y)\Big|.\end{aligned}$$

The first term in the right-hand side is bounded by

$$S_m \leq \frac{1}{m!}e^{-my}+\Big|S_m-\frac{1}{m!}e^{-my}\Big| \;;$$

the last term is less than $\frac{1}{m!}e^{-my}$. Thus in the interval $K^{-1} \leq F(y) \leq 1-K^{-1}$

$$\begin{aligned}\Delta(y) &\leq \frac{2}{m!}e^{-my}+\sum_{r=1}^{m}\Big|S_r-\frac{1}{r!}e^{-ry}\Big| \\ &\leq \frac{2}{m!}e^{-my}+C_7K^{-2}\sum_{r=1}^{m}\frac{1}{r!}e^{-ry} \\ &\leq \frac{2}{m!}e^{-my}+C_7K^{-2}F(y)^{-1} \\ &\leq \frac{2}{m!}e^{-my}+C_7K^{-1}.\end{aligned}$$

By the Stirling formula

$$\frac{1}{m!}e^{-my} \leq \left(\frac{e}{m}\right)^m e^{-my} \leq \left(\frac{e}{4}\right)^{4\log K} < K^{-1}.$$

Further, if $y<y_1$ where $F(y_1)=K^{-1}$,

$$\Delta(y) \le \max\{P(n^{-k}W(H_n)-\log|H_n|\le y),\ F(y)\}$$
$$\le \Delta(y_1)+2F(y_1)\le C_8K^{-1},$$

and similarly, for $y>y_2$ where $F(y_2)=1-K^{-1}$,

$$\Delta(y)\le \max\{P(n^{-k}W(H_n)-\log|H_n|>y),\ 1-F(y)\}$$
$$\le \Delta(y_2)+2(1-F(y_2))\le C_8K^{-1}.$$

Summarizing all these we obtain

$$\sup_y \Delta(y)\le C\,K^{-1},$$

which was to be proved.

## 4. PROOF OF (1)

Following the main lines of the above reasoning, put K=n. Since in the case k=1 the minimum of r waiting times is of geometrical distribution with parameter r/n, $S_r$ can be expressed explicitely:

$$S_r=\binom{n}{r}\left(1-\frac{r}{n}\right)^{[n(\log n+y)]}.$$

Hence we do not need $\varepsilon$. Suppose first that y is such that $n(\log n+y)$ is integer. Then

$$S_r=\frac{1}{r!}\exp\left(-\sum_{i=0}^{r-1}\log(n-i)+n(\log n+y)\log\left(1-\frac{r}{n}\right)\right).$$

If n is large enough, $m=[4\log n]<n/2$. Using the expansion

$$-\log(1-x)=x+\frac{1}{2}x^2+\ldots+\frac{1}{\ell}x^{\ell}+\theta\frac{1}{\ell+1}x^{\ell+1},\quad 1\le\theta<2,$$

valid for $0\le x\le 1/2$, we obtain the following expansion for the exponent in the above expression of $S_r$.

$$-\sum_{i=0}^{r-1}\log(n-i)+n(\log n+y)\log\left(1-\frac{r}{n}\right)$$
$$=-ry-r\sum_{j=1}^{2}n^{-j}P_j(r)-r(\log n+y)\sum_{j=1}^{2}n^{-j}Q_j(r)+$$
$$+O(n^{-3}\log^5 n),$$

where $P_j$ and $Q_j$ denote polynomials of degree $\le j$ and the remainder is uniform for $y$ belonging to the considered domain. Hence

$$S_r = \frac{1}{r!} e^{-ry}\left(1 - r\sum_{j=1}^{2}\sum_{i=0}^{j} n^{-j}(\log n + y)^i P_{ij}(r) + O(n^{-3}\log^9 n)\right),$$

where $\deg P_{ij} \le 2j-1$ and $P_{11}(r) = r$.

Polynomial $rP_{ij}(r)$ can be decomposed into a linear combination of the products $r(r-1)\ldots(r-t)$, $t=0,1,\ldots,2j-1$. After summation we get

$$\sum_{r=1}^{m-1}(-1)^r \frac{1}{r!} e^{-ry} rP_{ij}(r) = e^{-e^{-y}} e^{-y} Q_{ij}(e^{-y}) + O\left(\frac{1}{m!} e^{-my} m^{2j}\right),$$

where $\deg Q_{ij} \le 2j-1$ and $Q_{11}(z) = 1+z$. The first term in the right-hand side is a bounded function of $y$, and the remainder is $O(n^{-2}\log^{2j} n)$. Further,

$$\sum_{r=1}^{m}(-1)^r \frac{1}{r!} e^{-ry} O(n^{-3}\log^9 n) = e^{+e^{-y}} O(n^{-3}\log^9 n) = O(n^{-2}\log^9 n).$$

From all these we obtain that

$$P(n^{-1}W(\mathfrak{X}_n) - \log n \le y) = e^{-e^{-y}}\left(1 - e^{-y}(1+e^{-y})\frac{\log n + y}{n}\right) + O(1/n). \tag{12}$$

Dropping the condition that $n(\log n + y)$ should be integer, we only have to replace $y$ by $y + O(1/n)$ which yields an additional term $O(1/n)$ in (12). Finally, (12) can also be extended for $y$ with large modulus in the same way as in the proof of the Theorem. From (12) it follows that

$$\sup_y \left| P(n^{-1}W(\mathfrak{X}_n) - \log n \le y) - e^{-e^{-y}} \right| = e^{-c} c(1+c)\frac{\log n}{n} + O\left(\frac{1}{n}\right),$$

where $c = (\sqrt{5}+1)/2$, thus the proof of (1) is completed.

## REFERENCES

[1] P. Erdős, A. Rényi, 'On a classical problem of probability theory' *MTA Mat.Kut.Int.Közl.* 6A (1961) 215-220.

[2] S. R. Li, 'A martingale approach to the study of occurence of sequence patterns in repeated experiments' *Ann. Probab.* **8** (1980) 1171-1176.

[3] T. F. Móri, 'Large deviation results for waiting times in repeated experiments' *Acta Math. Hungar.* **45** (1985) 213-221.

[4] T. F. Móri, 'On the expectation of the maximum waiting time' *Ann. Univ.Sci.Bud.Eötvös, Sect.Comput.,* **7** to appear .

[5] T. F. Móri, 'On the waiting time till each of some given patterns occurs as a run' *Probab. Theory Appl.* to appear .

# ON BAYES INFERENCE IN CONTINGENCY TABLES

Zuzana Prášková, Monika Ratajová
Charles University, Department of Statistics
Sokolovská 83
186 00 Prague
Czechoslovakia

ABSTRACT. The multinomial sampling model for IxJ contingency table is considered. Assuming that the multinomial probabilities have certain prior distribution and utilizing the fact that the posterior distribution of the logarithmic interactions is asymptotically normal we construct credible intervals and simultaneous credible intervals for logarithmic interactions. The results are compared with those obtained by classical methods.

## 1. INTRODUCTION

Let us consider an IxJ contingency table with the counts $x_{ij} > 0$, $i=1,\ldots,I$, $j=1,\ldots,J$ without fixed marginals, having a multinomial distribution with the parameters $p_{ij} > 0$, $i=1,\ldots,I$, $j=1,\ldots,J$. Put $n = \Sigma\Sigma\, x_{ij}$ .

Let $\alpha = (\alpha_{ij})$ be a real matrix IxJ such that

$$\sum_{i=1}^{I} \alpha_{ij} = 0,\ j=1,\ldots,J, \qquad \sum_{j=1}^{J} \alpha_{ij} = 0,\ i=1,\ldots,I$$

and

$$\sum_{i=1}^{I} \sum_{j=1}^{J} |\alpha_{ij}| > 0 .$$

The interaction $\beta_\alpha$ corresponding to the matrix $\alpha$ defined by

$$(1) \qquad \beta_\alpha = \prod_{i=1}^{I} \prod_{j=1}^{J} p_{ij}^{\alpha_{ij}}$$

is a generalization of the cross-product ratio (odds-ratio)



P. Bauer et al. (eds.), Mathematical Statistics and Probability Theory, Vol. B, 179–188.

$\beta = p_{11}p_{22}/p_{12}p_{21}$ which is the well-known measure of association in a 2x2 table (in such a case we choose $\alpha_{11} = \alpha_{22} = 1$, $\alpha_{12} = \alpha_{21} = -1$).

The logarithmic interaction corresponding to $\alpha$ is defined by

$$(2) \qquad \delta_\alpha = \ln \beta_\alpha = \sum_{i=1}^{I} \sum_{j=1}^{J} \alpha_{ij} \ln p_{ij} \ .$$

It can be easily shown that under the hypothesis of independence, $\delta_\alpha = 0$ holds for any $\alpha$.

The sample interaction

$$b_\alpha = \prod_{i=1}^{I} \prod_{j=1}^{J} (x_{ij}/n)^{\alpha_{ij}} = \prod_{i=1}^{I} \prod_{j=1}^{J} x_{ij}^{\alpha_{ij}}$$

and the sample logarithmic interaction

$$(3) \qquad d_\alpha = \sum_{i=1}^{I} \sum_{j=1}^{J} \alpha_{ij} \ln (x_{ij}/n) = \sum_{i=1}^{I} \sum_{j=1}^{J} \alpha_{ij} \ln x_{ij}$$

are maximum likelihood estimators of $\beta_\alpha$ and $\delta_\alpha$, respectively.

All introduced definitions can be extended to any k-dimensional case. Let $(x_{i_1 \ldots i_k})$ be a k-dimensional contingency table with positive elements having a multinomial distribution with the parameters $p_{i_1 \ldots i_k} > 0$ where $i_j = 1,\ldots,I_j$, $j=1,\ldots,k$. Let $n$ denote the number of observations. Consider a k-dimensional array of real numbers $\alpha = (\alpha_{i_1 \ldots i_k})$ such that

$$\sum_{i_j=1}^{I_j} \alpha_{i_1 \ldots i_k} = 0 \qquad \text{for } j = 1,\ldots,k \ .$$

Then

$$\beta_\alpha = \Pi \ldots \Pi p_{i_1} \ldots p_{i_k}^{\alpha_{i_1 \ldots i_k}}$$

is called interaction corresponding to $\alpha$.

If we only suppose that

$$\Sigma \ldots \Sigma \alpha_{i_1 \ldots i_k} = 0 \quad \text{and} \quad \Sigma \ldots \Sigma |\alpha_{i_1 \ldots i_k}| > 0$$

then $\beta_\alpha$ is called generalized interaction corresponding to $\alpha$.

Logarithmic, respectively generalized logarithmic interactions $\delta_\alpha$ and their sample values are defined analogously. Notice, that the coefficient of partial association $p_{111}p_{221}/p_{121}p_{211}$ in a 2x2x2 table can be considered as a generalized interaction. In this case we put $\alpha_{111} = \alpha_{221} = 1$, $\alpha_{121} = \alpha_{211} = -1$, $\alpha_{112} = \alpha_{122} = \alpha_{212} = \alpha_{222} = 0$.

Provided that the multinomial model holds, the distribution of $\sqrt{n}(\hat{d}_\alpha - \delta_\alpha)$ is asymptotically normal if $n \to \infty$. Goodman (1964) and Anděl (1973) established simultaneous bounds for $\delta_\alpha$. In this section we shall assume that the parameters of multinomial distribution are random variables with given prior distribution and we shall construct simultaneous credible intervals for $\delta_\alpha$ based on the asymptotic posterior distribution of the multinomial parameters. Our results will be compared with those obtained by the classical confidence approach. We shall consider two-way contingency tables only but the results will be transferable to any k-variate case.

## 2. POSTERIOR DISTRIBUTION OF $\delta_\alpha$ AND BAYES BOUNDS

Let $X_1,\dots,X_t$ be random variables with the multinomial distribution with random parameters $p_1,\dots,p_t$. It is natural to suppose that the prior distribution of $p_1,\dots,p_t$ is Dirichlet with the parameters $\theta_1,\dots,\theta_t$. The posterior distribution given $x_1,\dots,x_t$ is Dirichlet, too, with the parameters $\theta_1+x_1,\dots,\theta_t+x_t$. For reasons discussed in Lindley (1964) we can confine ourselves to the limiting case $\theta_i = 0$, $i = 1,\dots,t$. Thus we shall assume that the joint prior distribution of $p_1,\dots,p_t$ is proportional to $\Pi\, p_i^{-1}$. The following lemma is only more exactly formulated Theorem 1 of Lindley (1964) and thus we give it without proof.

<u>Lemma 1</u>. Consider a sequence $X^\nu = (X_1^\nu,\dots,X_t^\nu)'$ of multinomial vectors with the parameters $n^\nu$, $p_1^\nu,\dots,p_t^\nu$, $p_i^\nu > 0$, $i = 1,\dots,t$, $\Sigma\, p_i^\nu = 1$. Suppose that the prior distribution of the vector $(p_1^\nu,\dots,p_t^\nu)'$ is proportional to $\Pi\,(p_i^\nu)^{-1}$. Let $A = (a_{ki})$, $k = 1,\dots,m$, $i = 1,\dots,t$ be a real matrix of the rank $m > t$ satisfying the conditions

$$\sum_{i=1}^{t} a_{ki} = 0 \ , \qquad \sum_{i=1}^{t} |a_{ki}| > 0, \quad k = 1,\ldots,m \ .$$

Let $x_1^\nu,\ldots,x_t^\nu$ be observed values of $X_1^\nu,\ldots,X_t^\nu$ , all positive, let $x_i^\nu = f_i n^\nu$ , $i = 1,\ldots,t$. Denote

$$P^\nu = (\ln p_1^\nu,\ldots,\ln p_t^\nu)'$$

$$L^\nu = (\ln x_1^\nu,\ldots,\ln x_t^\nu)' \quad .$$

Suppose that $n^\nu \to\infty$ if $\nu\to\infty$ ,f's fixed. Then the posterior distribution of $A\,P^\nu$ satisfies the condition

$$(4) \qquad \sqrt{n^\nu}(AP^\nu - AL^\nu) \longrightarrow N_m(0,AF^{-2}A') \quad \text{for} \quad \nu\to\infty$$

where

$$F^2 = \mathrm{Diag}[f_1,\ldots,f_t] \quad .$$

Having in mind the model described in Lemma 1 we shall omit the superscript $\nu$ for the sake of simplicity. Obviously, $x_i^\nu \to\infty$ if $\nu\to\infty$ for all $i = 1,\ldots,t$.

Let us consider an IxJ table with observed frequencies $(x_{ij})$ .Let $\delta_\alpha$ and $d_\alpha$ denote logarithmic, respectively sample logarithmic interaction corresponding to a matrix $\alpha = (\alpha_{ij})$. Put

$$S_{d_\alpha}^2 = \sum_{i=1}^{I} \sum_{j=1}^{J} \alpha_{ij}^2 \, x_{ij}^{-1} \quad .$$

If we suppose that the prior distribution of $(p_{11},\ldots,p_{IJ})'$ is proportional to $\Pi\Pi\, p_{ij}^{-1}$ , we obtain the following consequence of Lemma 1 for logarithmic interactions corresponding to matrices $\alpha_1,\ldots,\alpha_m$ .

Lemma 2. Provided the prior distribution of $(p_{ij})$ is proportional to $\Pi\Pi\, p_{ij}^{-1}$ , the posterior distribution of the vector

$$\left( \frac{\delta_1 - d_1}{S_{d_1}} \ ,\ldots,\ \frac{\delta_m - d_m}{S_{d_m}} \right)'$$

is asymptotically m-variate normal distribution with zero mean vector and unit diagonal elements of the dispersion matrix. The assertion holds true for generalized logarithmic interactions, too.

Proof. Put $f_{ij} = x_{ij}/n$ , $i = 1,\dots,I$, $j = 1,\dots,J$ ,

$$P = (\ln p_{11},\dots,\ln p_{IJ})'$$

$$L = (\ln x_{11},\dots,\ln x_{IJ})'$$

$$F^2 = \mathrm{Diag}\,[f_{11},\dots,f_{IJ}]$$

$$A = \begin{Vmatrix} a_1' \\ \vdots \\ a_m' \end{Vmatrix} = \begin{Vmatrix} \alpha_{11}^1 & ,\dots, & \alpha_{IJ}^1 \\ \vdots & & \\ \alpha_{11}^m & ,\dots, & \alpha_{IJ}^m \end{Vmatrix}$$

where $a_i'$ consists of elements $\alpha_{11}^i ,\dots,\alpha_{IJ}^i$ of the matrix $\alpha_i$ . Then the statement follows from (4) . Q.E.D.

Lemma 2 enables us to test hypotheses or construct credible intervals for logarithmic interactions. For instance, $(d_\alpha - u(1-\frac{1}{2}\gamma)s_{d_\alpha} , d_\alpha + u(1-\frac{1}{2}\gamma)s_{d_\alpha})$ is a credible interval for $\delta_\alpha$ with asymptotic credibility $1-\gamma$ ; $u(1-\frac{1}{2}\gamma)$ is $1-\frac{1}{2}\gamma$ - quantile of $N(0,1)$. (See Winkler(1972), Chapter 7.) As we shall see later, simultaneous Bayes bounds for a few logarithmic interactions can be constructed on the basis of Lemma 2. First we shall construct simultaneous credible intervals for all logarithmic interactions. We need some lemmas.

Lemma 3. Let $A(m\times t)$ and $F(t\times t)$ be as in Lemma 1 . Then $(AF^{-2}A')^{-1}$ is positive definite.

Proof. The matrix $W = AF^{-2}A'$ is regular and positive semidefinite which implies that it is positive definite. Then for any nonzero $x \in R_m$ we have $W^{-1}x = y \neq 0_m$ and

$$x'W^{-1}x = x'W^{-1}WW^{-1}x = (W^{-1}x)'W(W^{-1}x) = y'Wy > 0 \text{ .} \qquad \text{Q.E.D.}$$

Lemma 4. Let the assumptions of Lemma 1 be fulfilled. Then the posterior distribution of random variable

$$(5) \qquad n^\nu(P^\nu - L^\nu)'A'(AF^{-2}A')^{-1}A(P^\nu - L^\nu)$$

converges to $\chi^2$-distribution with $m$ degrees of freedom provided $\nu\to\infty$ .

Proof. Obviously, (5) is a continuous function of $\sqrt{n^\nu}A(P^\nu - L^\nu)$. Then the proof follows easily from (4) and from 2.c.4, XII, Rao (1973). Q.E.D.

Lemma 5. Let $M(m\times m)$ be a positive definite matrix. Then for any vector $x \in R_m$ the following equivalence holds:

$$[x'Mx \leq 1] \Longleftrightarrow [(h'x)^2 \leq h'M^{-1}h \quad \text{for all} \quad h \in R_m] \quad . \tag{6}$$

Proof. See Scheffè (1959), Appendix III.

Theorem 1. Consider an $I\times J$ table with the multinomial distribution having probabilities $p_{ij}$. Let the prior distribution of $(p_{ij})$ is proportional to $\Pi\Pi\, p_{ij}^{-1}$, let $(x_{ij})$ be observed frequencies. Let $\mathcal{A}$ be a set of all matrices $\alpha = (\alpha_{ij})$ such that

$$\sum_{i=1}^{I} \alpha_{ij} = \sum_{j=1}^{J} \alpha_{ij} = 0, \qquad \sum_{i=1}^{I} \sum_{j=1}^{J} |\alpha_{ij}| > 0 \quad .$$

Then, on the assumptions of Lemma 1, the posterior probability

$$P(|\delta_\alpha - d_\alpha| \leq \chi_m(1-\gamma)s_{d_\alpha} \qquad \text{for all } \alpha \in \mathcal{A})$$

is asymptotically (for $x_{ij} \to \infty$) equal to $1-\gamma$, where $m = (I-1)(J-1)$, $\chi^2_m(1-\gamma)$ is $(1-\gamma)$-quantile of $\chi^2$-distribution with $m$ degrees of freedom.

Proof. Write $\alpha \in \mathcal{A}$ as an IJ-dimensional vector. Let $\mathcal{L}(\mathcal{A})$ be the linear space over $\mathcal{A}$; obviously $\mathcal{L}(\mathcal{A}) = \mathcal{A} \cup \{0_m\}$. The dimension of $\mathcal{L}(\mathcal{A})$ is $m = (I-1)(J-1)$. Let $a_1^*,\dots,a_m^*$ be a basis of $\mathcal{L}(\mathcal{A})$. Put $A' = (a_1^*,\dots,a_m^*)$ and define $P$, $L$, $F^2$ similarly as in the proof of Lemma 2. Denote

$$\delta^* = AP \; , \quad d^* = AL \; , \quad W = AF^{-2}A' \; .$$

Then, according to Lemma 4, the posterior probability

$$P\,(n(\delta^* - d^*)'W^{-1}(\delta^* - d^*) \leq \chi^2_m(1-\gamma))$$

is asymptotically $1-\gamma$. From Lemma 3 we get that $W^{-1}$ is positive definite and thus we can use (6) to conclude that the posterior probability

$$P(|h'(\delta^* - d^*)| \leq \chi_m(1-\gamma)h'Wh/n \qquad \text{for all} \quad h \in R_m)$$

is asymptotically $1-\gamma$.

Let $\alpha \in \mathcal{A}$, then there exists $h \in R_m$ such that $h \neq 0_m$ and $\alpha = A'h$. Thus

$$h'\delta^* = h'AP = \alpha'P = \Sigma\Sigma\ \alpha_{ij}\ln p_{ij} = \delta_\alpha$$

$$h'd^* = h'AL = \alpha'L = \Sigma\Sigma\ \alpha_{ij}\ln x_{ij} = d_\alpha$$

$$h'Wh = h'AF^{-2}A'h = \alpha'F^{-2}\alpha = n\Sigma\Sigma\alpha_{ij}^2 x_{ij}^{-1} = nS_{d_\alpha}^2 \quad .$$

If $h$ varies over $R_m - \{0_m\}$ , $\alpha$ varies over $\mathcal{A}$ . The case $h = 0_m$ is trivial so that we can conclude that

$$P(|\ \delta_\alpha - d_\alpha| \leq \chi_m(1-\gamma)S_{d_\alpha} \quad \text{for all } \alpha \in \mathcal{A})$$

is asymptotically $1-\gamma$ . Q.E.D.

Theorem 1 gives simultaneous credible intervals

(7) $$(d_\alpha - \chi_{(I-1)(J-1)}(1-\gamma)S_{d_\alpha}\ ,\ d_\alpha + \chi_{(I-1)(J-1)}(1-\gamma)S_{d_\alpha})$$

for all logarithmic interactions . Remark that a similar result can be obtained for any k-variate case and for generalized interactions, too. Then the dimension of $\mathcal{L}(\mathcal{A})$ will be $m = (I_1-1)\ldots(I_k-1)$ for logarithmic interactions and $m = I_1\ldots I_k -1$ for generalized logarithmic interactions.

Bounds of credible intervals (7) could be rather wide if we consider only a few logarithmic interactions . In such a case we can utilize Lemma 2 for construction of appropriate Bayesian bounds.

<u>Theorem 2</u>. Let the assumptions of Lemma 1 be fulfilled. Let $\delta_k$, $d_k$, $S_{d_k}^2$ correspond to the matrices $\alpha_k$ , $k = 1,\ldots,r$. Then the posterior probability

(8) $$P(|\delta_k - d_k| \leq u(1-(2r)^{-1}\gamma)S_{d_k} \quad \text{for } k = 1,\ldots,r\ )$$

is asymptotically at least $1-\gamma$ ,where $u(1-\gamma)$ is $(1-\gamma)$-quantile of $N(0,1)$ .

Proof. According to Lemma 2 the posterior probability

$$P(|\delta_k - d_k| \leq u(1-(2r)^{-1}\gamma)S_{d_k}) \longrightarrow 1-\gamma r^{-1}$$

for $k = 1,\ldots,r$. Thus (8) follows from Bonferroni inequality. Q.E.D.

<u>Lemma 6.</u> If random variables $Y_1,\ldots,Y_k$ have a r-variate normal distribution with zero means, then for any positive $c_1,\ldots,c_r$

$$P(|Y_1| < c_1,\ldots,|Y_r| < c_r) \geq P(|Y_1| < c_1)\ldots P(|Y_r|<c_r).$$

Proof. See Šidák (1967).

Theorem 3. On the assumptions of Theorem 2 the posterior probability

$$P(|\delta_k - d_k| \leq u(\tfrac{1}{2}(1+(1-\gamma)^{1/r}))s_{d_k} \quad \text{for } k = 1,\dots,r) \tag{9}$$

is asymptotically at least $1-\gamma$.

Proof. According to Lemma 2 the joint posterior distribution of the vector

$$((\delta_1 - d_1)s_{d_1}^{-1},\dots,(\delta_r - d_r)s_{d_r}^{-1})'$$

converges to the distribution of a vector $(Y_1,\dots,Y_r)'$ which is normal with zero means and unit variances ($Y_i$, $Y_j$ for $i \neq j$ can be correlated.) Since

$$P(|Y_k| < u(\tfrac{1}{2}(1+(1-\gamma)^{1/r}))) = (1-\gamma)^{1/r}$$

for $k = 1,\dots,r$, the proof follows from Lemma 6.

Remark. Lemma 1 and Theorems 1-3 hold true also under the assumption $x_i^\nu/n^\nu \to f_i$, $i = 1,\dots,t$. It is easy to show that (4) remains valid on this assumption. Instead of (5) it is necessary to prove that the random variable $n^\nu(P^\nu-L^\nu)'A'(AF_\nu^{-2}A')^{-1}A(P^\nu-L^\nu)$ has asymptotically $\chi^2$-distribution with $m$ degrees of freedom, where

$$F_\nu^2 = \mathrm{Diag}\,[x_1^\nu/n^\nu,\dots,x_t^\nu/n^\nu] .$$

Obviously, $(AF_\nu^{-2}A')^{-1} \longrightarrow (AF^{-2}A')$. Similarly as in Lemma 3, $(AF_\nu^{-2}A')^{-1}$ is positive definite. Then there exists a regular matrix $H^\nu$ such that $H^\nu H^\nu = (AF_\nu^{-2}A')^{-1}$ and $\sqrt{n}H^\nu A(P^\nu-L^\nu) \longrightarrow N_m(0,1)$, from which our assertion follows.

Under the assumption that the prior distribution of the cell probabilities in a contingency table is limiting Dirichlet, the results of Theorems 1-3 closely coincide with those obtained by classical confidence approach (see Goodman (1964), Anděl (1973)). In Anděl's paper it is also shown that the bounds (9) are better than those given by (8) (see also Havránek (1978)).

Now let us assume that the prior distribution of the vector $(p_1^\nu,\dots,p_t^\nu)'$ considered in Lemma 1 is Dirichlet with the fixed parameters $\theta_1,\ldots,\theta_t$ . Then, under the assumption $x_i^\nu/n^\nu \to f_i$, (4) holds true with $L^\nu$ given by

$$L^\nu = (\ln (x_1^\nu + \theta_1),\ldots,\ln (x_t^\nu + \theta_t))'.$$

Thus, if we suppose that the prior distribution of $(p_{ij})$ is Dirichlet with fixed parameters $(\theta_{ij})$, the assertions of Theorems 1-3 remain valid with $d_\alpha$ and $S_{d_\alpha}$ in which the values $x_{ij}$ are replaced by $x_{ij}+\theta_{ij}$ for all i,j .

Example. The data in Table 1 are quoted in Plackett (1974). They show the relationship between nasal carrier rate for "Streptococcus pyogenes" and size of tonsils (in a natural ordering) among 1398 children age 0-15 years.

TABLE 1. Relationship between nasal carrier rate for Streptococcus pyogenes and size of tonsils

| | Present, but not enlarged | Enlarged | |
|---|---|---|---|
| | + | ++ | +++ |
| Carriers | 19 | 29 | 24 |
| Non-carriers | 497 | 560 | 269 |

The Pearson chi-square test for independence of the size of tonsils and the carrier rate yields $\chi^2 = 9.44770$ with 2 degrees of freedom which exceeds the 0.05 level of significance. For more detailed study of an association in this table we can consider matrices

$$\alpha_1 = \begin{Vmatrix} 1 & -1 & 0 \\ -1 & 1 & 0 \end{Vmatrix} \qquad \alpha_2 = \begin{Vmatrix} 1 & 0 & -1 \\ -1 & 0 & 1 \end{Vmatrix} \qquad \alpha_3 = \begin{Vmatrix} 0 & 1 & -1 \\ 0 & -1 & 1 \end{Vmatrix}$$

and construct simultaneous credible intervals $I_i$ for logarithmic interactions $\delta_i$ corresponding to $\alpha_i$, $i = 1,2,3$. We shall suppose that the prior distribution of the cell probabilities $p_{ij}$ is uniform on the set

$$\{p_{ij} > 0, \Sigma\Sigma p_{ij} = 1\}$$

i.e. Dirichlet with the parameters $\theta_{ij} = 1$ for all i,j . As $\chi_2(0.95) = 2.44775$, $u(1-0.05/6) = 2.39545$, and $u((1+(0.95)^{1/3})/2) = 2.38671$, we shall proceed according to Theorem 3 with $x_{ij}$ replaced by $x_{ij} + 1$ for all $i,j$. Thus, for $\gamma = 0.05$ we get

$$d_1 = -0.28634 \qquad I_1 = (-0.99082 \,;\, 0.41813)$$
$$d_2 = -0.83532 \qquad I_2 = (-1.57370 \,;\, -0.09694)$$
$$d_3 = -0.54898 \qquad I_3 = (-1.21904 \,;\, 0.12109)$$

The classical confidence approach yields

$$d_1 = -0.30351 \qquad I_1 = (-1.02314 \,;\, 0.41612)$$
$$d_2 = -0.84749 \qquad I_2 = (-1.60234 \,;\, -0.09254)$$
$$d_3 = -0.54398 \qquad I_3 = (-1.22598 \,;\, 0.13802)$$

We can see that the Bayesian intervals are shorter. In both cases the results indicate the association between carriers and the second category of enlarged tonsils but not among other categories.

## REFERENCES


1. Anděl, J. (1973). 'On Interactions in Contingency Tables.' Apl. mat. 18, 99-109.

2. Goodman, L.A. (1964). 'Simultaneous Confidence Limits for Cross-product Ratios in Contingency Tables.' J.Roy.Stat. Soc. ser. B, 26, 86-102.

3. Havránek, T. (1978). 'On Simultaneous Inference in Multidimensional Contingency Tables.' Apl. mat. 23, 31-38.

4. Lindley, D.V. (1964). 'The Bayesian Analysis of Contingency Tables' Ann. Math. Statist. 35, 1622-1643.

5. Plackett, R.L. (1974). The Analysis of Categorial Data. Griffin & Co., London.

6. Rao, R.C. (1973). Linear Statistical Inference and its Applications. Wiley, New York.

7. Scheffè, H. (1959). The Analysis of Variance. Wiley, New York.

8. Šidák, Z. (1967). 'Rectangular Confidence Regions for the Means of the Multivariate Normal Distribution.' J. Amer. Stat. Assoc. 62, 626-633.

9. Winkler, R.L. (1972). Introduction to Bayesian Inference and Decision. Holt, Rinehart and Winston, New York .

# SEQUENTIAL ESTIMATION FUNCTIONS IN STOCHASTIC POPULATION PROCESSES

Helmut Pruscha
Institute of Mathematics, University of Munich,
Theresienstr. 39
D-8000 Munich 2
West Germany

ABSTRACT. Sequential, i.e., randomly stopped estimation functions in a class of continuous time stochastic population models are considered. The class includes finite, irreducible Markov processes. Three types of efficient sequential estimation functions are discussed and their asymptotic behaviour is investigated. The main tools of analysis are taken from point process theory.

Some key words: Population processes, Markov branching process with immigration, birth-and-death-process, sequential estimation, Cramér-Rao inequality, asymptotic behaviour.

## 1. INTRODUCTION

The present paper deals with continuous time stochastic population models where the population is protected against extinction, possibly by an immigration component. The formulation of our model is broad enough to cover continuous time, finite state, irreducible Markov processes; in particular, Markov branching processes and birth-and-death processes are included in our approach. In fact, we use the concept of multivariate point processes, as was already earlier done by Aalen (1976), Johansen (1981), Jacobsen (1982), Pruscha (1985) and others. We are interested in those sequential, i.e., randomly stopped estimation procedures for the unknown parameters which are efficient in the Cramér-Rao sense. To this end, a multivariate Cramér-Rao inequality allowing a stopping time is established. Then three types of efficient sequential estimators are analyzed and their asymptotic properties (consistency, asymptotic normality) are investigated. Finally, special processes, which are covered by the approach, are discussed. Results by Magiera (1984), Franz (1982), Adke & Manjunath (1984) and others are generalized and new results on population processes where the Markovian framework is left are gained.

Note that Basawa & Becker (1983) recently employed stopping time procedures in stochastic population models to establish optimal inference for unknown parameters. Our main tools of analysis are taken from



*F. Bauer et al. (eds.), Mathematical Statistics and Probability Theory, Vol. B, 189–203.*

the theory of multivariate point processes.

## 2. MULTIVARIATE POINT PROCESS MODEL

First we introduce our stochastic model. Let $I$ be a finite set, $|I| = m$, and let

$$N_t = (N_{i,t}, i \in I), t \geq 0,$$

be a multivariate counting process on a probability space $(\Omega, \mathcal{F}, \mathbb{P})$; see Bremaud (1981) or Jacobsen (1982) for a survey.
Denote by $\mathcal{F}_t = \sigma(N_s, s \leq t), t \geq 0$, the internal history of the process. We assume that there exists an intensity process

$$\lambda_t = (\lambda_{i,t}, i \in I), t \geq 0,$$

which is nonnegative, adapted to $\mathcal{F}_t$ and left-continuous. Putting

$$A_{i,t} = \int_0^t \lambda_{i,s}\, ds$$

then

$$m_t = (m_{i,t}, i \in I), t \geq 0, \quad m_{i,t} = N_{i,t} - A_{i,t},$$

forms a local martingale with respect to $\mathcal{F}_t$, $t \geq 0$. To ensure non-extinction and non-explosion of the process, we assume throughout the paper that for all $i \in I$

A $\qquad A_{i,t} < \infty$ and $A_{i,t} \uparrow \infty \; (t \uparrow \infty)$ $\mathbb{P}$-almost sure.

Under this condition the following important properties of $N_t$ and $A_t$ hold true for all $i,j \in I$, where $\tau$ denotes a stopping time with respect to $\mathcal{F}_t$

(1) $\qquad \mathbb{E}\, N_{i,\tau} = \mathbb{E}\, A_{i,\tau}$

(2) $\qquad \mathbb{E}\, m_{i,\tau} m_{j,\tau} = \delta_{ij}\; \mathbb{E}\, A_{i,\tau} \qquad$ (if $\mathbb{E}\, A_{i,\tau} < \infty$)

(3) $\qquad N_{i,t}/A_{i,t} \to 1$ $\mathbb{P}$-almost sure $\quad (t \to \infty)$.

(1) and (2) can be found in Boel et al (1975) or Gill (1980) and (3) was proved by Lepingle (1978). If we assume that in probability $A_{i,t}/k_t \to a_i$ for each $i$, where $a_i > 0$ is a constant and $k_t$ a real function with $k_t \uparrow \infty$ as $t \uparrow \infty$, and if we define $Y_t = (Y_{i,t}, i \in I), t \geq 0$, by $Y_{i,t} = (N_{i,t} - A_{i,t})/\sqrt{A_{i,t}}$, then Aalen

(1976, p. 65) showed that under the condition $\mathbb{E}\, A_{i,t} < \infty$ in law

(4) $$Y_t \longrightarrow N_m \quad (t \to \infty)$$

where $N_m$ denotes the m-variate normal law with zero mean vector and unit covariance matrix.

Now let us assume in what follows that the intensity process $\lambda_t$, $t \geq 0$, can be represented in the scaled form

$$\lambda_{i,t} = \lambda_i(\theta)\varphi_{i,t}$$

where $\lambda_i(\theta)$ is positive and a function of the (unknown) parameter $\theta \in \Theta$, $\Theta \subset \mathbb{R}^d$ open. Putting $\Phi_{i,t} = \int_0^t \varphi_{i,s}\, ds$ condition A amounts to $\Phi_{i,t} < \infty$ and $\Phi_{i,t} \uparrow \infty$ as $t \uparrow \infty$ $\mathbb{P}_\theta$-almost sure.

B $\lambda_i(\theta)$ has continuous first-order derivatives with respect to $\theta \in \Theta$.

B allows to introduce for each $i$ and $\theta = (\theta_1,\ldots,\theta_d) \in \Theta$ the d-vector

$$f_i = \frac{d}{d\theta}\lambda_i(\theta)/\lambda_i(\theta)$$

and the $d \times m$ matrix F, possessing $f_i$, $i \in I$, as column vectors.

## 3. CRAMÉR-RAO INEQUALITY

Let $\tau$ be a stopping time with respect to $\mathcal{F}_t$, fulfilling

C $\tau < \infty$ $\mathbb{P}_\theta$- and $\Pi$-almost sure

where $\Pi$ stands for the probability law of $m$ independent homogeneous Poisson processes. Letting $\Pi_\tau$, $\mathbb{P}_{\theta,\tau}$ be the restriction of $\Pi$, $\mathbb{P}_\theta$ to $\mathcal{F}_t$, we have $\mathbb{P}_{\theta,\tau} \ll \Pi_\tau$ and with $l_\tau(\theta) = \log z_\tau(\theta)$, $z_\tau(\theta) = d\mathbb{P}_{\theta,\tau}/d\Pi_\tau$, and with $R_\tau$ not depending on $\theta$

(5) $$l_\tau(\theta) = \sum_{i\in I}\left(N_{i,\tau}\log\lambda_i(\theta) - \lambda_i(\theta)\Phi_{i,\tau}\right) + R_\tau,$$

see Kabanov et al (1976). We derive from (5) that the score function $J_\tau(\theta) = d\, l_\tau(\theta)/d\theta$ can be written as

(6) $$J_\tau(\theta) = F \cdot m_\tau \;.$$

As a consequence of (1) and (2) we have, if $\mathbb{E}_\theta\ \Phi_{i,\tau} < \infty$,

(7) $$\mathbb{E}_\theta\ J_\tau(\theta) = 0, \qquad \mathbb{E}_\theta\ J_\tau(\theta) J'(\theta) = I_\tau(\theta)$$

with the $d \times d$ matrix

$$I_\tau(\theta) = F \cdot \mathrm{Diag}\left(\mathbb{E}_\theta A_{i,t}(\theta)\right) \cdot F' \;.$$

We stipulate that

$D_1$ $\quad 0 < \mathbb{E}_\theta\ \Phi^2_{i,\tau} < \infty \quad$ for all $\ i \in I$

$D_2$ $\quad I_\tau(\theta)$ is nonsingular .

Note that $D_2$ is equivalent with $0 < \mathbb{E}\ \Phi_{i,\tau} < \infty$ for all i and rank (F) = d and that

$$\mathbb{E}_\theta\ N^2_{i,\tau} \leqq 2\left(\mathbb{E}_\theta\, m^2_{i,\tau} + \mathbb{E}_\theta A^2_{i,\tau}\right) < \infty$$

due to (2) and $D_1$. For some integer $q \geqq 1$ consider an q-dimensiomal process

$$g_t = \left(g_t^{(1)}, \ldots, g_t^{(q)}\right), \; t \geqq 0,$$

which is supposed to be right-continuous and adapted to $\mathcal{F}_t$ and which will play the role of an estimator. We introduce the conditions

$E_1$ $\quad \mathbb{E}_\theta \left(g_\tau^{(j)}\right)^2 < \infty$ for each j and $\mathbb{E}_\theta\, g_\tau$ has continuous derivatives with respect to $\theta$.

$E_2$ $\quad \lambda_i(\theta)$ is increasing (or decreasing) in the j-th component $\theta_j$ of $\theta$ for each $i,j$.

It can be proved that conditions A - E ensure that

(8) $$\frac{d}{d\theta}\ \mathbb{E}_\theta\ g'_\tau = \mathbb{E}_\Pi\ \frac{d}{d\theta}\ z_\tau(\theta) g'_\tau \quad ,$$

see Pruscha (1984, Th. 5.3.). Denote the left-hand side of (8) by $B_\tau(\theta)$. Using equations (7) and (8) the following theorem can be proved in the traditional way; $A \geqq 0$ means that $A$ is positive semidefinite.

<u>Theorem 1</u> Let assumptions A - E be fulfilled. Then

$$\mathrm{cov}_\theta(g_\tau) - B_\tau'(\theta) I_\tau^{-1}(\theta) B_\tau(\theta) \geqq 0 \tag{9}$$

with equality in (9) if and only if

$$g_\tau = \mathbb{E}_\theta g_\tau + B_\tau'(\theta) I_\tau^{-1}(\theta) J_\tau(\theta) \qquad \mathbb{P}_\theta\text{-almost sure.} \tag{10}$$

In the usual way one also derives

<u>Proposition 2</u> Under assumption A - E we have equality in (9) if and only if

$$g_\tau = c + C \cdot F \cdot m_\tau \qquad \mathbb{P}_\theta\text{-almost sure,} \tag{11}$$

where the q-vector $c$ and the $q \times d$ matrix $C$ do not depend on $\omega \in \Omega$.

## 4. SEQUENTIAL ESTIMATION FUNCTIONS

In what follows the basic conditions A, B are supposed to hold. Let us call the pair $(g,\tau)$, where $g_t$, $t \geqq 0$, is a right-continuous and $\mathcal{F}_t$-adapted process and $\tau$ an $\mathcal{F}_t$-stopping time, a <u>sequential</u> estimation function or <u>plan</u>. It is called <u>efficient</u> for the parameter function $h(\theta) = \mathbb{E}_\theta g_\tau$, if $\tau < \infty$ $\mathbb{P}_\theta$-almost sure and if (11) is fulfilled, for all $\theta \in \Theta$.

Now consider the special situation where

F $\qquad \mathrm{rank}(F) = m.$

<u>Proposition 3</u> Assume that A and B are true as well as $\tau < \infty$ $\mathbb{P}_\theta$-almost sure. The familiy of estimators of the form (11) is identical to the family of estimators of the form

$$g_\tau = c + D \cdot m_\tau \ , \tag{12}$$

$c$ q-vector, $D$ $q \times m$ matrix, both not depending on $\omega$, if condition F holds. If the m-vectors $m_\tau$ span the whole $\mathbb{R}^m$ for varying $\omega$ then the converse is also true.

<u>Proof</u> There is for each $q \times m$ matrix $D$ a $q \times d$ matrix $C$ with $C \cdot F = D$, if and only if $\mathrm{rank}(F) = m$.

If $\mathbb{E}_\theta \Phi_{i,\tau} < \infty$ then (12) implies that $\mathbb{E}_\theta g_\tau = c$ and

$\mathrm{cov}_\theta(g_\tau) = D \cdot \mathrm{Diag}\big(\mathbb{E}\, A_{i,\tau}(\theta)\big) \cdot D'$ as a consequence of (1) and (2). The advantage of (12) over (11) is that the matrix F is no longer involved. With respect to (11) or (12) it will be sufficient to consider the case $q = 1$, that is to say, one-dimensional estimation functions.

## 5. THREE TYPES OF EFFICIENT PLANS

First let us rewrite the stochastic model in the following manner: Divide the finite set I into M disjoint subsets $I^{(j)}$, $j = 1,\dots, M$ $(M \leqq m)$, such that $\Phi_{i,t}$ is the same for all $i \in I^{(j)}$. Put

$$\Phi_t^{(j)} = \Phi_{i,t} \quad , \qquad i \in I^{(j)}$$

and introduce double indices in $N_t$ an $\lambda_t$; rewrite $N_t$ as

$$N_t = \big(N_{i,t}^{(j)}\, , i \in I^{(j)}\, , j = 1,\dots,M\big)$$

and $\lambda$ in the same way. The dependency of $\lambda_t$ on $\theta$ will be suppressed in the following. Let us fix some j, j = 1 say, and some $k \leqq k_1 = |I^{(1)}|$. Put $I^{(1)} = \{1,\dots,k_1\}$ and $N_{+,t}^{(1)} = \sum_{i=1}^{k} N_{i,t}^{(1)}$ , $\lambda_+^{(1)} = \sum_{i=1}^{k} \lambda_i^{(1)}$ .

For some integer $n_1 \geqq 1$ we define the stopping time with respect to $\mathcal{F}_t$

$$\tau^{(1)} = \inf\,\{t \geqq 0 : N_{+,t}^{(1)} = n_1\} \;.$$

Note that condition A and (3) ensure that $\tau^{(1)} < \infty$ $\mathbb{P}_\theta$-almost sure. For some real $t_1 > 0$ we define a second stopping time with respect to $\mathcal{F}_t$

$$\tau^{(2)} = \inf\,\{t \geqq 0 : \Phi_t^{(1)} = t_1\} \;,$$

and have $\tau^{(2)} < \infty$ $\mathbb{P}_\theta$-almost sure from A. A third stopping time with respect to $\mathcal{F}_t$

$$\tau^{(3)} = \inf\,\{t \geqq 0 : N_{+,t}^{(1)} = a\,\Phi_t^{(1)} - b\}$$

with $a > 0$ is introduced as well as the condition

G $\qquad \lambda_+^{(1)} < a \quad$ for all $\theta \in \Theta$ .

Lemma 4 Let condition A be fulfilled as well as G and $b > 0$ in the case $i = 3$. We have for $\tau = \tau^{(i)}$, $i = 1,2,3$,

$$\mathbb{E}_\theta \left(N_{+,\tau}^{(1)}\right)^2 < \infty, \qquad \mathbb{E}_\theta \left(\Phi_\tau^{(1)}\right)^2 < \infty. \tag{13}$$

Proof Let $N = N_+^{(1)}$, $\Phi = \Phi^{(1)}$. Introducing the random time change

$$\sigma(s) = \inf \{t \geq 0 : \Phi_t \geq s\}, \; N_s^* = N_{\sigma(s)} ,$$

we know that the counting process $N_s^*$, $s \geq 0$, is distributed with respect to the $\mathbb{P}_\theta$-probability like a homogeneous Poisson process with parameter $\lambda_+^{(1)}$. Treating the case $\tau = \tau^{(3)}$ only, whe show that

$$\tau = \sigma(\tau^*), \qquad \tau^* = \inf \{s \geq 0 : N_s^* = as - b\}.$$

In fact, since $N_{t_1} = N_{t_2}$ $\mathbb{P}_\theta$-almost sure if $\Phi_{t_1} = \Phi_{t_2}$ $(t_1 < t_2)$ by Jacobsen (1982, Prop. 2.2.10c) or Brémaud (1981, Th.II 12) we have

$$N_{\Phi_t}^* = N_{\sigma(\Phi_t)} = N_t ,$$

and therefore

$$\tau = \inf \{t \geq 0 : N_{\Phi_t}^* = a\,\Phi_t - b\} = \sigma(\tau^*) .$$

Now, $\mathbb{E}_\theta (\tau^*)^2 < \infty$ by Trybula (1968, p.23) from where

$$\mathbb{E}_\theta (\Phi_\tau)^2 = \mathbb{E}_\theta (\Phi_{\sigma(\tau^*)})^2 = \mathbb{E}_\theta (\tau^*)^2 < \infty$$

due to the continuity of $\Phi_t$, $t \geq 0$ .

Remark Let $\tau = \tau^{(3)}$ and $b \leq 0$; then it follows from Dwass (1984, p. 359, 368) that

$$\mathbb{P}_\theta (\tau = \infty) \geq \mathbb{P}_\theta (N_s^* < as - b \text{ for all } s > 0) = 1 - \lambda_+^{(1)}/a > 0.$$

To formulate the following result we introduce the notation

$$f^{(1)} = \lambda_+^{(1)}, \quad f^{(2)} = 1, \quad f^{(3)} = a - \lambda_+^{(1)}$$

$$r^{(1)} = n_1, \quad r^{(2)} = t_1, \quad r^{(3)} = b \ .$$

Proposition 5 Let conditions A,B,F be fulfilled as well as G and $b > 0$ in the case $i = 3$, and let $c_0,\dots,c_{k_1}$ be arbitrary real numbers.

Then the parametric function

$$(14) \qquad h^{(i)}(\theta) = \left(c_0 + \sum_{j=1}^{k_1} c_j \lambda_j^{(1)}\right)/f^{(i)}$$

is efficiently estimated by $g_{\tau^{(i)}}/r^{(i)}$, $i = 1,2,3$, where

$$(15) \qquad g_t = c_0 \Phi_t^{(1)} + \sum_{j=1}^{k_1} c_j N_{j,t}^{(1)} \ .$$

Proof In each of the three cases we are now going to rewrite $g_{\tau^{(i)}}$ in the form (12).

Case i = 1: $(\tau = \tau^{(1)})$ $g_\tau$ can be written as

$$(16) \qquad g_\tau = n_1 h^{(1)}(\theta) + \sum_{j=1}^{k_1} \alpha_j \left(N_{j,\tau}^{(1)} - \lambda_j^{(1)}\Phi_\tau^{(1)}\right)$$

with $\alpha_j = c_j - h^{(1)}(\theta)$ for $j = 1,\dots,k$, and $\alpha_j = c_j$ for $j = k+1,\dots,k_1$.

Case i = 2: $(\tau = \tau^{(2)})$ $g_\tau$ can be written as

$$(17) \qquad g_\tau = t_1 h^{(2)}(\theta) + \sum_{j=1}^{k_1} c_j \left(N_{j,\tau}^{(1)} - \lambda_j^{(1)}\Phi_\tau^{(1)}\right).$$

Case i = 3: $(\tau = \tau^{(3)})$ $g_\tau$ can be written as

$$g_\tau = \sum_{j=1}^{k_1} c_j \left(N_{j,\tau}^{(1)} - \lambda_j^{(1)}\Phi_\tau^{(1)}\right) + \left(c_0 + \sum_{j=1}^{k_1} c_j \lambda_j^{(1)}\right)\Phi_\tau^{(1)} \ .$$

The second summand equals

$$h^{(3)}(\theta)\left((a\,\Phi_\tau^{(1)} - b) - \lambda_+^{(1)}\Phi_\tau^{(1)}\right) + bh^{(3)}(\theta) =$$

$$= h^{(3)}(\theta) \sum_{i=1}^{k} \left(N_{i,\tau}^{(1)} - \lambda_i^{(1)}\Phi_\tau^{(1)}\right) + bh^{(3)}(\theta) \ .$$

Hence

$$g_\tau = bh^{(3)}(\theta) + \sum_{j=1}^{k_1} \beta_j\left(N_{j,\tau}^{(1)} - \lambda_j^{(1)}\Phi_\tau^{(1)}\right) \tag{18}$$

with $\beta_j = c_j + h^{(3)}(\theta)$ for $j = 1,\dots,k$, and $\beta_j = c_j$ for $j = k+1,\dots,k_1$.
In all three cases (13) and (1) yield $\mathbb{E}_\theta g_{\tau^{(i)}} = r^{(i)}h^{(i)}(\theta)$, while an application of Prop. 3 finishes the proof.

Remarks (i) Similar methods of proof were already used by Bai (1975), Adke & Manjunath (1984), Manjunath (1984) in special cases. Note that Prop. 5 also follows from Stefanov (1986), where a different method is applied.

(ii) Special choices of the $c_i$'s, as $c_0 = 1$, $c_j = 0$ else or $c_{j_1} = 1$ $(j_1 \neq 0)$, $c_j = 0$ else, yield various parametric functions of interest. The case $i = 3$ can be generalized by replacing $N_{+,t}^{(1)}$ and $\lambda_+^{(1)}$ by $\sum \alpha_j N_{j,t}^{(1)}$ and $\sum \alpha_j \lambda_j^{(1)}$, resp., where the nonnegative $\alpha_j$ are not all zero, see Trybula (1982), Stefanov (1984).

## 6. ASYMPTOTIC RESULTS

The notation of the last section remains valid in the following. The stopping times $\tau = \tau^{(i)}$ can be indexed by the value of $r = r^{(i)}$, such that we are led to write $\tau_r$. We have in all three cases $i = 1,2,3$ with $b > 0$ in case 3,

$$\tau_r \to \infty \quad \text{as } r \to \infty \quad \mathbb{P}_\theta\text{-almost sure,} \tag{19}$$

since $N_{+,t}^{(1)} < \infty$, $\Phi_t^{(1)} < \infty$ for each $t \geqq 0$ $\mathbb{P}_\theta$-almost sure by A and (3). We first prove the consistency of the efficient sequential estimators given in proposition 5.

Theorem 6 Let condition A be fulfilled as well as G and $b > 0$ in the case $i = 3$. Consider the estimator $g_t$ and the parametric functions $h^{(i)}(\theta)$, $i = 1,2,3$, as given in (15) and (14), resp., and let $\tau = \tau^{(i)}$, $r = r^{(i)}$. Then we have as $r \to \infty$

$$g_{\tau_r}/r \longrightarrow h^{(i)}(\theta) \qquad \mathbb{P}_\theta\text{-almost sure } (i = 1,2,3).$$

Proof We can write

$$g_{\tau_r} = \left(c_0 + \sum_{j=1}^{k_1} c_j \left(N^{(1)}_{j,\tau_r}/\Phi^{(1)}_{\tau_r}\right)\right) \Phi^{(1)}_{\tau_r} .$$

Since

$$(20) \qquad N^{(1)}_{+,\tau^{(1)}_n} = n, \quad \Phi^{(1)}_{\tau^{(2)}_t} = t, \quad a\Phi^{(1)}_{\tau^{(3)}_b} - N^{(1)}_{+,\tau^{(3)}_b} = b$$

we deduce from (3) that $\mathbb{P}_\theta$-almost sure

$$(21) \qquad \Phi^{(1)}_{\tau_r}/r \longrightarrow 1/f^{(i)}, \quad i = 1,2,3, \quad N^{(1)}_{j,\tau_r}/\Phi^{(1)}_{\tau_r} \longrightarrow \lambda^{(1)}_j$$

as $r \to \infty$. Thus the theorem is proved.

For the proof of the asymptotic normality of the efficient sequential estimators we have to replace condition A by a stronger one

$$A^* \qquad \mathbb{E}_\theta\, \Phi^{(i)}_t < \infty \text{ and } \Phi^{(i)}_t/t \longrightarrow g^2 \ (t \to \infty) \ \mathbb{P}_\theta\text{-almost sure}$$

for each $i,\theta$, where $g^2$ is some positive constant.

We deduce from $A^*$, (20) and (3) that for $\tau = \tau^{(i)}$, $r = r^{(i)}$, $i = 1,2,3$,

$$(22) \qquad \tau_r/r \longrightarrow 1/\left(g^2 f^{(i)}\right) \quad \mathbb{P}_\theta\text{-almost sure}$$

as $r \to \infty$. Next we establish the asymptotic normality of efficient sequential estimators.

Theorem 7 Let condition $A^*$ be fulfilled as well as G and $b > 0$ in the case $i = 3$. Consider the estimator $g_t$ and the parametric function $h^{(i)}(\theta)$, $i = 1,2,3$, as given in (15) and (14), resp., and let $\tau = \tau^{(i)}$, $r = r^{(i)}$. Then we have in the $\mathbb{P}_\theta$-law

$$\sqrt{r}\left(g_{\tau_r}/r - h^{(i)}(\theta)\right) \longrightarrow N\left(0, \sum_{j=1}^{k_1} \gamma_j^2\right)$$

as $r \to \infty$, where

$$\gamma_j^2 = \begin{cases} \alpha_j^2 \, \lambda_j^{(1)}/\lambda_+^{(1)} & , \quad \text{case } i = 1 \\ c_j^2 \, \lambda_j^{(1)} & , \quad \text{case } i = 2 \\ \beta_j^2 \, \lambda_j^{(1)}/\left(a - \lambda_+^{(1)}\right) & , \quad \text{case } i = 3 \end{cases}$$

and $\alpha_j, \beta_j$ are given by (16) and (18), resp. We also have joint distributional convergence w.r. to all components $j = 1,\ldots,M$ (see the beginning of section 5).

Proof Let us summarize equations (16) - (18) by

$$g_{\tau_r} = r\, h^{(i)}(\theta) + \sum_{j=1}^{k_1} \eta_j \, m_{j,\tau_r}^{(1)} ,$$

where $\eta_j = \alpha_j$, $c_j$ and $\beta_j$ in the three cases $i = 1,2$ and 3, resp., and where $m_{j,t}^{(1)} = N_{j,t}^{(1)} - A_{j,t}^{(1)}$, $A_{j,t}^{(1)} = \lambda_j^{(1)} \Phi_t^{(1)}$. Now

$$\sqrt{r}\left(g_{\tau_r}/r - h^{(i)}(\theta)\right) = \sum_{j=1}^{k_1} \eta_j \, m_{j,\tau_r}^{(1)} / \sqrt{r} =$$

$$= \sum_{j=1}^{k_1} \eta_j \left(A_{j,\tau_r}^{(1)}/r\right)^{\frac{1}{2}} m_{j,\tau_r}^{(1)} \Big/ \left(A_{j,\tau_r}^{(1)}\right)^{\frac{1}{2}} .$$

By virtue of (21) and (4), resp., we have as $r \to \infty$ and $t \to \infty$, resp.

$$A_{j,\tau_r}^{(1)}/r \longrightarrow \lambda_j^{(1)}/f^{(1)}, \quad i = 1,2,3, \quad \mathbb{P}_\theta\text{-almost sure}$$

$$\Sigma_j \, \eta_j \, m_{j,t}^{(1)} \Big/ \left(A_{j,t}^{(1)}\right)^{\frac{1}{2}} \longrightarrow N\left(0, \, \Sigma_j \, \eta_j^2\right) \quad \text{in } \mathbb{P}_\theta\text{-law.}$$

To finish the proof, we have to use the random time change theorem of Billingsley (1968, Th.17.1) which is applicable due to (22), and to note, that (4) also yields joint distributional convergence.

Remark Note that the results of Stefanov (1986) enable another proof of Th. 7 as far as one component ($j = 1$, say) is considered and that Sørensen (1986) contains some exact (nonasymptotic) distributional results.

# 7. SPECIAL PROCESSES

## 1. Continuous time, finite state Markov processes

If the state space of the Markov process $X_t$ consists of the states $1,\ldots, m$ and if $(q_{ij})$ is the infinitesimal $m \times m$ intensity matrix, we put

$$I^{(i)} = \{(i,j),\ \ j = 1,\ldots,m\},\ \ i = 1,\ldots,m \quad ,$$

and

$$\lambda_{j,t}^{(i)} = q_{ij}\ 1(X_{t-} = i),\ \ i \neq j,\ \ \lambda_{i,t}^{(i)} = 0 \quad .$$

Hence

$$\lambda_j^{(1)} = q_{1j},\ \ j \neq 1,\ \ \Phi_t^{(1)} = \int_0^t 1(X_s = 1)ds,$$

$\Phi_t^{(1)}$ being the total time in state 1 up to time t. Thus, Prop. 5 contains results of Adke & Manjunath (1984) and Trybula (1982). If the Markov process is irreducible and positive recurrent, the condition A* is guaranteed, see Azema et al (1967, p.170).

## 2. Markov branching process with immigration

If $(p_j,\ 0 \leqq j \leqq m)$ denotes the offspring distribution with $p_1 = 0$, $(b_j,\ 1 \leqq j \leqq m)$ the immigrant distribution, $\lambda$ and $\mu\varphi_t^{(0)}$ the individual split- and overall immigration -rate, resp., then we put

$$I^{(i)} = \{(i,j),\ \ j = -1,1,\ldots,m\},$$

$$i = 0(\text{'immigration'}) \ \text{ or } = 1(\text{'split'}),$$

and

$$\lambda_{j,t}^{(i)} = \begin{cases} \mu b_j\ \varphi_t^{(0)} & \text{for } i = 0 \\ \\ \lambda p_{j+1}\ X_{t-} & \text{for } i = 1 \end{cases} ,$$

where $b_{-1} = 0$, $p_{m+1} = 0$, and

$$X_t = x_0 + \sum_{j=1}^m j\ N_{j,t}^{(0)} + \sum_{j=-1}^m j\ N_{j,t}^{(1)}$$

is the population size at time t. The process $\varphi_t^{(0)}$, $t \geqq 0$, is supposed to be left-continuous and adapted to $\mathcal{F}_t$ which leaves the Markovian framework. Put $\Phi_t^{(0)} = \int_0^t \varphi_s^{(0)}\, ds$ and ssume that $\mathbb{E}\, \Phi_t^{(0)} < \infty$. Condition $A^*$ is fulfilled if $\Sigma_j\, jp_j < 1$ and $\Phi_t^{(0)}/t \to h^2 > 0$ $\mathbb{P}$-almost sure as $t \to \infty$, see Pruscha (1985, L.4.12, Th.4.16), where asymptotic results in the non-sequential case can be found. Note that Franz (1982) treats the special case of a linear birth-and-death process with immigration and that other special cases are treated by Magiera (1984) and Jang & Bai (1986).

## 3. Nonlinear birth-and-death process with immigration

With $i = -1, 0$ and $1$ denoting the event of death, immigration and birth, resp., we define

$$\lambda_t^{(1)} = \beta X_{t-}, \quad \lambda_t^{(-1)} = \delta X_{t-}^2, \quad \lambda_t^{(0)} = \mu\varphi_t^{(0)}, \tag{23}$$

where $\beta, \delta$ and $\mu$ are positive constants, where the process $\varphi_t^{(0)}$, $t \geq 0$, is supposed to be left-continuous and adapted to $\mathcal{F}_t$, and where

$$X_t = x_0 + N_t^{(0)} + N_t^{(1)} - N_t^{(-1)}$$

is the population size at time t. Once again put $\Phi_t^{(0)} = \int_0^t \varphi_s^{(0)} ds$ and suppose that $\mathbb{E}\, \Phi_t^{(0)} < \infty$. We are going to discuss the condition $A^*$ for the birth component $i = 1$. Put $\Phi_t^{(1)} = \int_0^t X_s\, ds$. To establish $\mathbb{E}\, \Phi_t^{(1)} < \infty$, we consider the linear birth-and-death process $\tilde{X}_t$ with immigration defined by setting the right-hand sides of (23) as $\beta\tilde{X}_{t-}$, $\delta\tilde{X}_{t-}$, $\mu\varphi_t^{(0)}$. We have $\mathbb{E}\, X_t \leq \tilde{\mathbb{E}}\, \tilde{X}_t$ by using a domination principle similar to that of Jacobsen (1982, p.31) and by applying formula (1) and $\tilde{\mathbb{E}} \int_0^t \tilde{X}_s\, ds < \infty$ by using Pruscha (1985, L.4.12). To establish $\Phi_t^{(1)}/t \to g^2$ we first prove

<u>Lemma 8</u> Consider the model (23) with $\mu = 0$. Then

$$K \equiv \mathbb{E} \int_0^\infty X_s\, ds < \infty .$$

<u>Proof</u> We have $\mathbb{E}\, X_t = \mathbb{E}\left(X_t | X_t > 0\right) \mathbb{P}\left(X_t > 0\right)$. By the exponential

ergodicity of the process ($c_i$ positive constants)

$$\mathbb{P}\left(X_t > 0\right) \leqq c_1 e^{-c_2 t}$$

since Th.5.3(ii) of Van Dorn (1985) can be used after relabeling the states from x into x - 1. As in subcritical linear birth-and-death processes we have $\mathbb{E}\left(X_t | X_t > 0\right) \leqq c_3$, see Harris (1963, Th.V.11.1), which completes the proof.

Now we are able to establish the second half of condition A* for i = 1 the proof of which is similar to that of Th.4.16 in Pruscha (1985).

<u>Theorem 9</u> If $\Phi_t^{(0)}/t \to h^2 > 0$ $\mathbb{P}$-almost sure as $t \to \infty$, then

$$\Phi_t^{(1)}/t \to g^2$$

$\mathbb{P}$-almost sure with the constant $g^2 = \mu h^2 K$.

<u>Acknowledgement</u> The author is grateful to a referee for some useful hints and remarks.

REFERENCES

AALEN, O.O. (1976). 'Statistical inference for a family of counting processes'. Inst. of Math. Statist., University of Copenhagen.

ADKE, S.R. & MANJUNATH, S.M. (1984). 'Sequential estimation for continuous time finite Markov processes.' *Commun. Statist. - Theor. Meth.* **13**, 1089-1106.

AZÉMA, J., KAPLAN-DUFLO, M. & REVUZ, D. (1967). 'Mésure invariante sur les classes récurrent des processus de Markov.' *Z. Wahrsch. theorie verw. Gebiete* **8**, 157-181.

BAI, D.S. (1975). 'Efficient estimation of transition probabilities in a Markov chain.' *Ann. Statist.* **6**, 1305-1317.

BASAWA, I.V. & BECKER, N. (1983). 'Remarks on optimal inference for Markov branching processes: a sequential approach.' *Austral. J. Statist.* **25**, 35-46.

BILLINGSLEY, P. (1968). *Convergence of Probability Measures*. Wiley, New York.

BOEL, R., VARAIYA, P. & WONG, E. (1975). 'Martingales on jump processes I.' *Siam J. Control* **13**, 999-1021.

BRÉMAUD, P. (1981). *Point Processes and Queues*. Springer-Verlag, Berlin

DWASS, M. (1974). 'Poisson process and distribution-free statistics.' *Adv. Appl. Prop.* **6**, 359-375.

FRANZ, J. (1982). 'Sequential estimation and asymptotic properties in birth-and-death processes.' *Math. Operationsforsch. Statist. ser. Statist.* **13**, 231-244

GILL, R.D. (1980). 'Censoring and stochastic integrals.' *Math. Centre Tracts* **124**, Math. Centrum Amsterdam.

HARRIS, T.E. (1963). *The Theory of Branching Processes*. Springer-Verlag Berlin.

JACOBSEN, M. (1982). 'Statistical Analysis of Counting Processes.' *Lecture Notes in Statistics* **12**, Springer-Verlag, New York.

JANG, J.S. & BAI, D.S. (1986). 'Efficient sequential estimation in Markov branching processes with immigration.' *Comm. Statist.-Theor. Meth.* **15**, 1681-1697.

JOHANSEN, S. (1981). 'The statistical analysis of Markov branching processes.' Inst. of Math. Statist., University of Copenhagen.

KABANOV, JU.M., LIPTSER, R.S. & SHIRYAYEV, A.N. (1976). 'Criteria of absolute continuity of measures corresponding to multivariate point processes.' In: *Lecture Notes in Math.* **550**, 232-252, Springer-Verlag, Berlin

LEPINGLE, D. (1978). 'Sur le comportement asymptotic des martingales locales.' In: *Lecture Notes in Math.* **649**, 148-161, Springer-Verlag, Berlin

MAGIERA, R. (1984). 'Sequential estimation of the tramsition intensities in Markov processes with migration.' *Zastosowania Matematyki* **18**, 241-250.

MANJUNATH, S.M. (1984). 'Optimal sequential estimation for ergodic birth-death processes.' *J. Roy Statist. Soc. B* **46**, 412-418.

PRUSCHA, H. (1984). 'Parametric inference in multivariate point processes.' Reprint, University of Munich.

PRUSCHA, H. (1985). 'Parametric inference in Markov branching processes with time-dependent random immigration rate.' *J. Appl. Prob.* **22**. 503-517.

SØRENSEN, M. (1986). 'On sequential maximum likelihood estimation for exponential families of stochastic processes.' *Int. Statist. Rev.* **54**, 191-210.

STEFANOV, V.T. (1984). 'Efficient sequential estimation in finite-state Markov processes.' *Stochastics* **11**, 291-300.

STEFANOV, V.T. (1986). 'Efficient sequential estimation in exponential type processes.' To appear in *Ann. Statist.*

TRYBULA, S. (1968). 'Sequential estimation in processes with independent increments.' *Dissertationes Mathematicae* **60**.

TRYBULA, S. (1982). 'Sequential estimation in finite-state Markov processes.' *Zastosowania Matematyki* **17**, 227-248.

VanDOORN, E.A. (1985). 'Conditions for exponential ergodicity and bounds for the decay parameter of a birth-death process.' *Adv. Appl. Prob.* **17**, 514-530.

## ESTIMATING QUADRATIC POLYNOMIALS WITH APPLICATIONS TO SQUARE ROOT NORMALIZING TRANSFORMATIONS

Andrew L. Rukhin
Department of Mathematics and Statistics
University of Massachusetts
Amherst, MA 01003
U.S.A.

ABSTRACT. Assume that the observed sample $y_1,\dots,y_n$ is approximately normalized by the square root transformation, $x = 2(y^{1/2}) - 1)$ which belongs to the Box-Cox family of normalizing transformations. The unknown mean of the original $y$'s sample is a quadratic function of the normal parameters of the transformed sample $x_1,\dots,x_n$. We study the behavior of a natural estimator of this and more general quadratic functions of normal parameters and obtain a necessary and sufficient condition for its admissibility under quadratic loss. In the case of inadmissibility, a class of better estimators is exhibited.

### 1. INTRODUCTION AND SUMMARY

Let $y_1,\dots,y_n$, $n \geq 2$, be a random sample of (positive) observations. Assume that this data is approximately normalized by the square root transformation, $x = 2(y^{1/2} - 1)$ which is a member of the Box-Cox family of normalizing transformations often used in practice (see Box and Cox (1964), Carroll and Ruppert (1984)). In this situation the mean of the $y$'s sample is $(\xi^2 + \sigma^2)/4 + \xi + 1$. Here $\xi$ and $\sigma$ denote unknown parameters of a normal distribution of the transformed sample $x_1,\dots,x_n$.

In this paper we consider the more general estimation problem of a quadratic function $\theta = \alpha\xi^2 + \beta\sigma^2 + \gamma\xi$ for given $\alpha$, $\beta$ and $\gamma$. This parametric function includes as particular cases, the variance $(\alpha = \gamma = 0)$, the logarithm of a log normal mean $(\alpha = 0,\ \beta = 1/2,\ \gamma = 1)$, the second moment $(\alpha = \beta = 1,\ \gamma = 0)$ and some interesting functions of $\xi$ and $\sigma$ corresponding for instance to more general normalizing square root type transformations. Without loss of generality we shall assume throughout this paper that $\alpha$ is nonnegative.

Let $X = \Sigma_1^n x_j/n$, $S^2 = \Sigma_1^n(x_j - X)^2$ be a version of the sufficient



*P. Bauer et al. (eds.), Mathematical Statistics and Probability Theory, Vol. B, 205–214.*

statistic for $\xi$ and $\sigma$. The best unbiased estimator of $\theta$ has the form

$$\delta_U(X,S) = \alpha X^2 + (\beta - \alpha n^{-1})S^2/(n-1) + \gamma X .$$

If $\beta \neq \alpha n^{-1}$ this estimator is inadmissible under quadratic loss. Indeed let us consider estimators of the form

$$\delta(X,S) = \alpha X^2 + cS^2 + \gamma X$$

where $c$ is a real constant.

To evaluate the mean squared error of this estimator we use the independence of $X$ and $S$ and the form of their first moments and obtain

$$\begin{aligned}(1.1)\qquad E(\delta(X,S)-\theta)^2 &= \alpha^2 E(X^2-\xi^2-\sigma^2/n) + c^2E(S^2-(n-1)\sigma^2)^2 \\ &+ \sigma^4(c(n-1)+\alpha/n-\beta)^2 + \gamma^2E(X-\xi)^2 \\ &+ 2\alpha\gamma E(X-\xi)(X^2-\xi^2-\sigma^2/n) \\ &= \sigma^2(2\alpha\xi+\gamma)^2/n + \sigma^4[2\alpha^2/n + c^2(n^2-1) \\ &+ 2c(n-1)(\alpha/n-\beta) + (\beta-\alpha/n)^2].\end{aligned}$$

Therefore there exists the optimal choice of the constant $c$

$$c = a = (\beta - \alpha/n)(n+1).$$

In this paper we study the admissibility of the resulting estimator

$$(1.2)\qquad \delta_0(X,S) = \alpha X^2 + aS^2 + \gamma X$$

under rescaled quadratic loss $(\delta - \theta)^2\sigma^{-4}$.

Estimator of (1.2) also improves on the maximum likelihood estimator $\hat{\delta}(X,S) = \alpha X^2 + \beta S^2/n + \gamma X$.

One can also consider more general quadratic estimators of the form $bX^2 + cS^2 + dX$ for some $b$, $c$ and $d$. However a calculation like in (1.1) shows that the choice $b = \alpha$, $c = a$, $d = \gamma$ is optimal for large values of $\xi/\sigma$.

Notice that $\delta_0$ can be represented in the following form

$$\begin{aligned}\delta_0(X,S) &= \alpha(X^2 - S^2/[n(n+1)]) + \beta S^2/(n+1) + \gamma X \\ &= \alpha\delta_1(X,S) + \beta\delta_2(X,S) + \gamma X.\end{aligned}$$

Clearly $\delta_1(X,S)$ is the estimator of $\xi^2$ which is optimal within the classes $x^2 + cS^2$, and $\delta_2(X,S)$ is the best equivariant estimator of $\sigma^2$ (i.e. is optimal in the class $cS^2$).

The inadmissibility of $\delta_2$ is well known (cf. Stein (1964), Brown (1968), Brewster and Zidek (1974), Strawderman (1974)). It was established that improvement over $\delta_2$ can be found in the class of shrinkage estimators $\delta$ (i.e. $0 \leq \delta \leq \delta_2$). Estimator $\delta_1$ of $\xi^2$ is also inadmissible, in fact $\max(\delta_1,0)$ is better than $\delta_1$. This improvement is an "expander", not a "shrinker". Thus the inadmissibility of $\delta_0$ cannot be concluded from its inadmissibility in the particular cases $\alpha = 0$, $\beta > 0$ and $\beta = 0$, $\alpha > 0$. As a matter of fact for "small" values of $\beta$, $\beta n > (n + 2)\alpha$, there are expanding statistics which improve on $\delta_0$, and for "large" values of $\beta$, $\beta n \geq (2n + 3)\alpha$, some shrinkage estimators are better than $\delta_0$. For "intermediate" values of $\beta$, $\alpha(n + 2) \leq \beta n < (2n + 3)\alpha$ the estimator $\delta_0$ is proven to be admissible. Heuristically for these values of $\beta$ the inadmissibilities of $\delta_1$ and $\delta_2$ "cancel" each other, and $\delta_0$ is a generalized Bayes rule against a prior which admits a good approximation in terms of posterior risk by proper priors. Thus the admissibility of $\delta_0$ in this case can be proven by standard technique (cf. Blyth (1951), Farrell (1969)). The results mentioned above are obtained by the reduction of the general case with $\alpha > 0$ to the situation when $\gamma = 0$ studied by the author (Rukhin (1987)).

In this paper we also determine a class of estimators which improve upon $\delta_0$ when $\beta n < \alpha(n + 2)$.

To summarize our results the following Table is provided.

TABLE 1. Admissibility of Estimator (1.2)

| Values of $\alpha,\beta$ ($\gamma$ arbitrary, $\geq 0$) | $\delta_0$ is admissible |
|---|---|
| $\alpha = \beta = 0$ | Yes |
| $\alpha = 0$, $\beta \neq 0$ | No |
| $(n+2)\alpha/n \leq \beta < (2n+3)\alpha/n$ | Yes |
| $\beta < (n+2)\alpha/n$ or $\beta \geq (2n+3)\alpha/n$ | No |

## 2. GENERALIZED BAYES ESTIMATORS

It is well known that a statistical procedure is admissible if and only if it admits a good approximation in terms of posterior risk by proper Bayes rules (cf. for instance Farrell (1968)). In this section we investigate the form of prior densities $\lambda$ for which $\delta_0$ is a formal Bayes estimator and find out which of these densities admit the approximation by proper priors.

Let $\lambda(\xi,\sigma)$ be the density of the generalized prior distribution over $\{(\xi,\sigma),\sigma > 0\}$ with respect to the uniform measure $d\xi d\sigma/\sigma$. (The latter is "noninformative" Jeffreys prior in a location-scale parameter model. It coincides with the right Haar measure over the corresponding group of linear transformations of the real line).

The Bayes estimator of $\theta = \alpha\xi^2 + \beta\sigma^2 + \gamma\xi$ has the form

$$\delta_B(x,s) = \frac{\int_{-\infty}^{\infty}\int_0^{\infty} \theta\sigma^{-n-4}\exp\{-[n(x-\xi)^2+s^2]/(2\sigma^2)\}\lambda(\xi,\sigma)d\xi d\sigma}{\int_{\infty}^{\infty}\int_0^{\infty} \sigma^{-n-4}\exp\{-[n(x-\xi)^2+s^2]/(2\sigma^2)\}\lambda(\xi,\sigma)d\xi d\sigma}.$$

Assume that $\lambda$ is sufficiently smooth, so that the following integrations by parts are legitimate:

(2.1) $$\int_{-\infty}^{\infty}(\xi-x)\lambda\ \exp\{-[n(x-\xi)^2+s^2]/(2\sigma^2)\}d\xi$$
$$= \sigma^2 n^{-1}\int_{-\infty}^{\infty}\lambda_\xi \exp\{-[n(x-\xi)^2+s^2]/(2\sigma^2)\}d\xi,$$

(2.2) $$\int_{-\infty}^{\infty}(\xi-x)\lambda\ \exp\{-[n(x-\xi)^2+s^2]/(2\sigma^2)\}d\xi$$
$$= \sigma^2 n^{-1}\int_{-\infty}^{\infty}[\lambda+\xi\lambda_\xi]\exp\{-[n(x-\xi)^2+s^2]/(2\sigma^2)\}d\xi,$$

(2.3) $$\int_{-\infty}^{\infty}(\xi-x)^2\lambda\exp\{-[n(x-\xi)^2+s^2]/(2\sigma^2)\}d\xi$$
$$= \sigma^2 n^{-1}\int_{-\infty}^{\infty}[\lambda+\sigma^2 n^{-1}\lambda_{\xi\xi}]\exp\{-[n(x-\xi)^2+s^2]/(2\sigma^2)\}d\xi,$$

(2.4) $$[s^2+n(x-\xi)^2]\int_0^{\infty}\lambda\sigma^{-n-5}\exp\{-[n(x-\xi)^2+s^2]/(2\sigma^2)\}d\xi$$
$$= \int_0^{\infty}[(n+2)\lambda-\sigma\lambda_\sigma]\sigma^{-n-3}\exp\{-[n(x-\xi)^2+s^2]/(2\sigma^2)\}d\sigma.$$

Combining these formulae we obtain the following representation of the Bayes estimator

(2.5) $$\delta_B(x,s) - \delta_0(x,s) = \int_{-\infty}^{\infty}\int^{\infty}[\mathcal{D}\lambda]\sigma^{-n-3}\exp\{-[n(x-\xi)^2+s^2]/2\sigma^2\}d\xi d\sigma$$
$$/n\int_{-\infty}^{\infty}\int_0^{\infty}\lambda\sigma^{-n-5}\exp\{-[n(x-\xi)^2+s^2]/2\sigma^2\}d\xi d\sigma$$

where

$$\mathcal{D}\lambda = (a - \alpha/n)\sigma^2\lambda_{\xi\xi} + (2\alpha\xi + \gamma)\lambda_\xi + an\sigma\lambda_\sigma + 2\alpha\lambda .$$

It follows from (2.5) that $\delta_B = \delta_0$ if and only if the density $\lambda$ is a solution of the following parabolic differential equation

$$\mathcal{D}\lambda = 0. \tag{2.6}$$

Thus we proved the following

THEOREM 1. Assume that the prior density $\lambda$ is differentiable and integrations by parts on (2.1)-(2.4) are legitimate. Then the formal Bayes estimator $\delta_B$ has form (2.5) and $\delta_B = \delta_0$ if and only if (2.6) holds.

Notice that (2.6) has an evident solution $\lambda(\xi,\sigma) = \sigma^{-2\alpha/an}$, but it also has many other solutions. To find them for $\alpha > 0$ put $\lambda(\xi,\sigma) = \psi(\xi + \gamma/(2\alpha),\sigma)$. Then

$$\mathcal{D}\psi = (a - \alpha/n)\sigma^2\psi_{\xi\xi} + 2\alpha\xi\psi_\xi + an\sigma\psi_\sigma + 2\alpha\psi ,$$

i.e. equation (2.6) reduces to the case $\gamma = 0$, studied in Rukhin (1987).

In particular, if

$$\psi_0(\xi,\sigma) = \exp\{-n(2\alpha - an)\xi^2][2(an - \alpha)\sigma^2]\}/\sigma \tag{2.7}$$

then

$$\mathcal{D}\psi_0 = 0.$$

If $\alpha < an < 2\alpha$, i.e. if $\alpha(n+2) < \beta n < \alpha(2n+3)$,

$$\int_{-\infty}^{\infty} \psi_0(\xi,\sigma)d\xi < \infty \tag{2.8}$$

for all $\sigma$.

If $an = \alpha$ the general solution of (2.6) has the form

$$\lambda(\xi,\sigma) = \sigma^{-2}f((\xi + \gamma/2\alpha)/\sigma^2) \tag{2.9}$$

with an arbitrary function $f$.

These facts suggest the admissibility of $\delta_0$ in the case $\alpha \le an < 2\alpha$, and in Section 3 we prove this result.

On the other hand because of the admissibility of $\delta_0$ when $\gamma = 0$, $\beta n < (n+2)\alpha$ or $\beta n \ge (2n+3)\alpha$, (Rukhin (1987)), equation (2.6) cannot have solutions which are approximable by proper densities for these values of $\beta$.

It is easy to see that if $\tilde{\delta}_B$ denotes the Bayes estimator of $\alpha\xi^2 + \beta\sigma^2$ against prior density $\psi(\xi,\sigma) = \lambda(\xi - \gamma/(2\alpha),\sigma)$ then

$$\tilde{\delta}_B(X,S) = \delta_B(X - \gamma/(2\alpha),S) + \gamma^2/(4\alpha^2) .$$

Thus for $\alpha > 0$ the Bayes estimators of $\theta$ and $\alpha\xi^2 + \beta\sigma^2$ are closely related.

If $\alpha = 0$, equation (2.6) takes the form

$$(2.10) \qquad (n+1)\mathcal{D}\lambda = \beta(\sigma^2\lambda_{\xi\xi} + (n+1)\lambda_\xi/\beta + n\sigma\lambda_\sigma) = 0.$$

Consider first the case when $\gamma = 0$, $\beta \neq 0$. Then (2.10) can be rewritten as

$$(2.11) \qquad \sigma^2\lambda_{\xi\xi} + n\sigma\lambda_\sigma = 0$$

and this equation which is closely related to the adjoint heat equation, does not have solutions approximable by proper densities because of the inadmissibility of $\delta_0$ as an estimator of $\sigma^2$.

If $\gamma \neq 0$, $\beta \neq 0$, put

$$\lambda(\xi,\sigma) = \mathcal{H}(\xi - \gamma(n+1)\log \sigma/(\beta n),\sigma).$$

Then

$$\sigma^2\mathcal{H}_{\xi\xi} + n\sigma\mathcal{H}_\sigma = 0$$

i.e. $\mathcal{H}$ must be a solution of (2.11). Thus in this case $\mathcal{H}$ (as well as $\lambda$) does not admit an approximation by proper prior densities and $\delta_0$ cannot be admissible. In fact the inadmissibility of $\delta_0$ was established in Rukhin (1986b).

At last if $\alpha = \beta = 0$, equation (2.10) reduces to

$$\lambda_\xi = 0$$

with general solution $\lambda(\xi,\sigma) = \lambda(\sigma)$. Some of these solutions can be approximated by proper priors. As a matter of fact, the admissibility of $\delta_0(X,S) = X$ as an estimator of $\xi$ is well known.

## 3. ADMISSIBILITY AND INADMISSIBILITY RESULTS

We start here with the formualtion of the main result.

THEOREM 2. For $\alpha > 0$, $\tilde{\delta}_0$ is an admissible estimator of $\theta = \alpha\xi^2 + \beta\sigma^2 + \gamma\xi$ under quadratic loss if and only if $(n+2)\alpha \leq \beta n < (2n+3)\alpha$. If $\alpha = 0$, $\beta \neq 0$, estimator $\delta_0$ is inadmissible, and if $\alpha = \beta = 0$ it is admissible.

Proof: Because of the discussion in Section 2 we consider the case $\alpha > 0$ only, and use the reduction to $\gamma = 0$.

Let $\tilde{\delta}(X,S)$ be any estimator of $\alpha\xi^2 + \beta\sigma^2$. Then for $\tau = \gamma/(2\alpha)$, $\mu = \xi + \tau$,

$$\begin{aligned} E_{\xi\sigma}(\tilde{\delta}(X,S) - \alpha\xi^2 - \beta\sigma^2)^2 &= E_{\mu\sigma}(\tilde{\delta}(X+\tau,S) - \alpha\xi^2 - \beta\sigma^2)^2 \\ &= E_{\mu\sigma}(\tilde{\delta}(X+\tau,S) - \alpha\tau^2 - \alpha\mu^2 - \beta\sigma^2 - \gamma\mu)^2. \end{aligned}$$

In other terms the risk function of $\tilde{\delta}$ at $(\xi,\sigma)$ coincides with that of

$$\delta(X,S) = \tilde{\delta}(X + \tau,S) - \alpha\tau^2$$

as an estimator of $\theta$ at $(\mu,\sigma)$. Therefore $\tilde{\delta}$ is admissible if and only if $\delta$ is. In particular, if $\tilde{\delta}_0(X,S) = \alpha X^2 + aS^2$,

$$\tilde{\delta}(X+\tau,S) - \alpha\tau^2 = \alpha X^2 + aS^2 + \gamma X = \delta_0(X,S)$$

and the statement above about risks can be obtained directly from (1.1).

By considering the posterior risk of Bayes estimators for densities approximating (2.7) and (2.9) and using (2.8) it is shown in Rukhin (1987) that $\tilde{\delta}_0$ is admissible if $\alpha(n+2) \leq \beta n < \alpha(2n+3)$. By constructing explicit alternative estimators it is also proven there that if $0 \leq \beta n < \alpha(n+2)$ or $\beta n \leq \alpha(2n+3)$, $n > 3$, then $\tilde{\delta}_0$ is inadmissible. A careful analysis of the proof shows that the same inadmissibility conclusion holds in the case $\beta < 0$.

It follows from Theorem 2 that the estimator $\delta_0$ of the mean of the pretransformed sample $(\alpha = \beta = 1/4, \gamma = 1)$ is inadmissible.

The better estimators given in Theorem 1 of Rukhin (1987) cannot be admissible when $\alpha(n+2) > \beta n$ because of the lack of regularity. Motivated by the case $\alpha = \beta$ when this inequality holds we derive here a class of smooth improvements some of which can be shown to be generalized Bayes estimators.

The estimators under consideration have the form (for $\gamma = 0$)

$$\delta(X,S) = \delta_0(X,S) + 2n^{-1}\rho S^2 h(U) \tag{3.1}$$

where $U = n^{1/2}|X|(nX^2 + S^2)^{-1/2}$, $\rho = \alpha - an$ and $h$ is a positive measurable function. These scale-equivariant estimators have been introduced by Stein (1964) and studied by Brown (1986), Brewster and Zidek (1974) and Strawderman (1974).

THEOREM 3. Assume that $\beta n < \alpha(n+2)$ and $\gamma = 0$. Also suppose that for some positive $p$, $h(u)(1-u^2)^{-p}$ is a nonincreasing function of $u$, $0 \leq u \leq 1$. Then the corresponding estimator for (3.1) improves on $\delta_0$ if

$$h(0) \leq \min[(p+1)^{-1}, r_p(n+2)^{-1}] \tag{3.2}$$

where

$$r_p = 2pB(3/2,(n+1)/2+p)/B(1/2,(n+3)/2+2p).$$

Proof: It is easy to see that the risk of any estimator (3.1) depends only on $\eta = n^{1/2}\xi/\sigma$, so that one can put $\sigma = 1$. Let

$$d_k = \int_0^\infty e^{-\tau^2/2}\tau^k dt = 2^{(k-1)/2}\Gamma((k+1)/2),$$

$$C = 8\rho(2\pi)^{-1/2}/[d_{n-2}\cdot n] .$$

Then

$$\begin{aligned}\Delta(\eta) &= E[(\delta_0-\theta)^2-(\delta-\theta)^2] \\ &= Ce^{-\eta^2/2}\int_0^1\int_0^\infty h(u)s^{n+1}[-\alpha s^2u^2/(1-u^2) - ans^2 - \rho s^2h(u) + \\ &\quad + \alpha\eta^2 + \beta n]\exp\{-s^2/2(1-u^2) + su\eta(1-u^2)^{-1/2}\}(1-u^2)^{-3/2}duds \\ &= Ce^{-\eta^2/2}\sum_{k=0}^{\infty} \eta^{2k}d_{n+2k+3}/(2k)! \\ &\quad \int_0^1 u^{2k}(1-u^2)^{(n-1)/2}h(u)\{\beta n/(n+2+2k) - \alpha u^2 - \\ &\quad - (an+\rho h(u))(1-u^2) + 2k(2k-1)\alpha/[u^2(n+2+2k)(n+2k)\}du .\end{aligned} \tag{3.3}$$

Thus $\Delta$ is nonnegative for all $\eta$ and $\delta$ is better than $\delta_0$ if for $k = 0,1,\ldots$ each integral on (3.3) is nonnegative. This condition holds if

$$\int_0^1 u^{2k}(1-u^2)^{(n-1)/2}h(u)\{\beta n/(n+2+2k) - u^2 - \tag{3.4}$$

$$-(an+ h(0)(1-u^2)^p)(1-u^2) + 2k(2k-1)\alpha/$$

$$[u^2(n+2+2k)(n+2)]\}du \geq 0 .$$

The expression in brackets in (3.4) changes sign at most once from positive in the interval [0,1] provided that

$$\rho h(0)(p+1) \leq \alpha - an = \rho$$

i.e. if $h(0) \leq (p+1)^{-1}$.

Now one can apply Baranchik's lemma (cf. Baranchik (1970), Strawderman (1974) or Rukhin (1986a)) to conclude that (3.4) holds if

$$\int_0^1 u^{2k}(1-u^2)^{(n-1)/_2+p}\{\beta n/(n+2+2k) - \alpha u^2 - an(1-u^2) - \tag{3.5}$$

$$- \rho h(0)(1-u^2)^{p+1} + 2k(2k-1)\alpha/[u^2(n+2+2k)(n+2k)]\}du \geq 0 .$$

The latter integral can be expressed in terms of beta function, so that (3.5) is met if

$$\rho h(0)B(k+1/2,(n+3)_{/2}+2p) \leq \beta nB(k+1/2,(n+1)_{/2}+p)/(n+2+2k) -$$

$$-\alpha B(k+3/2,(n+1)_{/2}+p) - anB(k+1/2,(n+3_{/2}+p) +$$

$$+ 4k\alpha B(k+1/2,(n+1)/_2+p((n_{/2}+p+k)/[(n+2+2k)(n+2k)].$$

Thus $h$ corresponds to a better estimator if

$$\rho h(0) \leq \Gamma((n+1)_{/2}+p)/\Gamma((n+3)_{/2}+2p)$$

$$\times \min_{k\geq 0}\{\beta n((n+2)_{/2}+p+k)/(n+2+2k) - (k+1/2) - an((n+1)_{/2}+p)+$$

$$+ 4k\alpha((n+2)_{/2}+p+k)(n_{/2}+p+k)[(n+2+2k)(n+2k]\}$$

$$\times \Gamma((n+4)_{/2}+2p+k)/\Gamma((n+4)_{/2}+p+k) = \rho r_p/(n+2)$$

i.e. if (3.2) is valid.

Notice that the bound (3.2) coincides with that in Theorem 1 of Strawderman (1974) for $\alpha = 1$. Thus our Theorem 3 generalizes the class of minimax estimators of unknown normal variance. Following Strawderman (1974) one can also obtain generalized Bayes estimators which satisfy the conditions of Theorem 3.

REFERNECES

Baranchik, A. J. (1970). 'A family of minimax estimators of the mean of a multivariate normal distribution'. Ann. Math. Statist. 41, 642-645.

Blyth, C. R. (1951). 'On minimax statistical decision procedures and their admissibility'. Ann. Math. Statist. 22, 22-42.

Box, G. E. P. and D. R. Cox (1964). 'An analysis of transformations'. Journal Royal Statist. Soc. Ser. B, 26, 211-252.

Brewster, J. F. and J. V. Zidek (1974). 'Improving on equivariant estimators'. Ann. Math. Statist. 2, 21-38.

Brown, L. (1986). 'Inadmissibility of the usual estimators of scale parameters'. Ann. Math. Statist. 39, 29-48.

Carroll, R. J. and D. Ruppert (1984). 'Power transformations when fitting theoretical models to data'. Journal Amer. Statist. Assoc. 79, 321-328.

Farrell, R. (1964). 'Estimators of a location parameter in the absolutely continuous case'. Ann. Math. Statist. 35, 949-998.

Farrell, R. (1968). 'On a necessary and sufficient condition for admissibility of estimators when strictly convex loss is used'. Ann. Math. Statist., 39, 23-28.

Rukhin, A. L. (1986a). 'Admissibility and minimaxity results in the estimation problem of exponential quantiles'. Ann. Math. Statist. 14, 220-237.

Rukhin, A. L. (1986b). 'Estimating a linear function of the normal mean and variance'. Sankhya, 48.

Rukhin, A. L. (1987). 'Quadratic estimators of quadratic functions of normal parameters'. J. Statist. Planning and Inference, 14.

Stein, C. (1964). 'Inadmissibility of the usual estimator for the variance of a normal distribution with unknown mean'. Ann. Inst. Statist. Math. 16, 155-160.

Strawderman, W. E. (1974). 'Minimax estimation of powers of the variance of a normal population under squared error loss'. Ann. Math. Statist. 2, 190-198.

# A $k$-SAMPLE PROBLEM WITH CENSORED DATA

A. Schick and V. Susarla
State University of New York at Binghamton
Department of Mathematical Sciences
Binghamton, New York 13901
U.S.A.

ABSTRACT. In this paper, we address the question of asymptotically efficient estimation in randomly right censored regression models. We allow the censoring distributions to depend on the covariate. For simplicity, we consider only situations in which the covariates come from a finite set. We provide a characterization of efficient estimates, describe a general method for the construction of such estimates and carry out this construction when the censoring distributions are known.

## 1. *Introduction*

In several statistical experiments pertaining to clinical/longitudinal investigations, the observations involve covariate values for the items under observation and the data on these items might be randomly right censored. This is commonly referred to as the randomly right censored regression model. The object of this paper is to obtain asymptotically efficient estimates for the regression parameter when the covariate values come from a finite set. By asymptotically efficient estimates, we mean it in the sense of local asymptotic minimax as expounded in Schick (1986b). It is stressed here that the censoring distribution is allowed to depend upon the covariate value. A specialization of the results presented here obtains the results of Schick, Susarla, and Koul (1986). As in this reference, we treat various cases depending upon whether the error distribution and/or censoring distributions is and/or are known. We show that not knowing the censoring distributions does not cause any loss of efficiency when compared to the situation in which they are known. On the contrary, not knowing the error distribution will result in some loss of efficiency. In other words, adaptive estimation in the sense of Begun, Hall, Huang, and Wellner (1983), Bickel (1982) and Fabian and Hannan (1982) is not possible. Finally, we will construct efficient estimates in the above models. The construction of these estimates uses the asymptotic results presented in Gill (1983), Schick (1986a), Schick, Susarla, and Koul (1986).

We now provide notation and definitions which are used throughout the rest of the paper. The (multivariate) normal distribution with mean vector $m$ and covariance matrix $V$ is denoted by $\mathcal{N}(m, V)$. Let $\lambda$ be the Lebesgue measure on the Borel $\sigma$-field of the real line $\mathbf{R}$ and $\mu$ be a $\sigma$-finite measure on the Borel $\sigma$-field of the extended real line $(-\infty, \infty]$. Define a $\sigma$-finite measure $\nu$ on the Borel $\sigma$-field $\mathbf{S}$ of $S = \{0,1\} \times \mathbf{R}$ by $\nu(A) = \lambda(A_0) + \mu(A_1)$,



P. Bauer et al. (eds.), Mathematical Statistics and Probability Theory, Vol. B, 215–230.

$A \in \mathbf{S}$ where $A_\delta = \{z \in \mathbf{R} : (\delta, z) \in A\}$ for $\delta = 0, 1$. Let $\mathcal{F}$ denote the class of Lebesgue densities and $\mathcal{G}$ be the class of $\mu$-densities. Define a function $p$ on $S \times \mathbf{R} \times \mathcal{F} \times \mathcal{G}$ by

$$p((\delta, z) \mid t, \phi, \gamma) = \delta\phi(z - t) \int_{(z,\infty]} \gamma d\mu + (1 - \delta)\gamma(z) \int_{z-t}^{\infty} \phi d\lambda. \tag{1.1}$$

Then $p(\cdot \mid t, \phi, \gamma)$ is a $\nu$-density. A random vector which has this density occurs in random right censoring models in which the life-time variable has $\lambda$-density $\phi(\cdot - t)$ and the independent censoring variable has $\mu$-density $\gamma$. For a positive integer $k$, $(\Delta_{ij}, Z_{ij})$, $i = 1, ..., k$, $j = 1, 2, ...$, are $S$-valued random vectors and, for each $t \in \mathbf{R}^k$ and $u = (\phi, \gamma_1, ..., \gamma_k) \in \mathcal{F} \times \mathcal{G}^k$, $P_{t,u}$ is a probability measure for which the random vectors $(\Delta_{ij}, Z_{ij})$, $i = 1, ..., k$, $j = 1, 2, ...$ are independent and $(\Delta_{ij}, Z_{ij})$ has $\nu$-density $p(\cdot \mid t_i, \phi, \gamma_i)$. Fix $\theta = (\theta_1, ..., \theta_k) \in \mathbf{R}^k$ and $\eta = (f, g_1, ..., g_k) \in \mathcal{F} \times \mathcal{G}^k$. Let $F = \int_\cdot^\infty f d\lambda$, $G_i = \int_{(\cdot,\infty]} g_i d\mu$, and $p_i = p(\cdot \mid \theta_i, f, g_i)$, $i = 1, ..., k$. In the following $\int_a^b$ stands for $\int_{(a,b)}$.

Our problem is to estimate $\theta$ efficiently based on $\{(\Delta_{ij}, Z_{ij} : j = 1, ..., n_i, i = 1, ..., k\}$ if they are generated by $P_{\theta,\eta}$ and

$$n_i \to \infty \tag{A.1}$$

and

$$n_i/n \to \lambda_i > 0 \text{ where } n = n_1 + ... + n_k. \tag{A.2}$$

The following regularity conditions on $(\theta, \eta)$ will be imposed throughout.

$$\int F(\cdot - \theta_i) g_i d\mu < 1 \text{ for } i = 1, ..., k, \tag{A.3}$$

and

$$f \text{ is absolutely continuous and } J(f) = \int (f')^2/f d\lambda < \infty. \tag{A.4}$$

We briefly discuss now some implications of (A.1) - (A.4). For this, let $I$ denote the $k \times k$ diagonal non-singular matrix with $i$th diagonal entry given by

$$I_i = \lambda_i \left\{ \int \frac{f'^2}{f}(\cdot - \theta_i) G_i d\lambda + \int \frac{f^2}{F}(\cdot - \theta_i) g_i d\mu \right\} \tag{1.2}$$

Conditions (A.1) - (A.4) imply the local asymptotic normality of the parametric model at $\theta$, i.e.,

$$\left\{ \begin{array}{l} \text{for every bounded sequence } t_n \in \mathbf{R}^k, \\ \Lambda_n(t_n) - \Lambda_n(0) - t_n^T \gamma_n - \frac{1}{2} t_n^T I t_n^T \to 0, \text{ in } P_{\theta,\eta}\text{-prob.} \end{array} \right. \tag{1.3}$$

and

$$\mathcal{L}(\gamma_n \mid P_{\theta,\eta}) \Longrightarrow \mathcal{N}(0, I) \tag{1.4}$$

where

$$\Lambda_n(t) = \sum_{i=1}^{k}\sum_{j=1}^{n_i} \log p(\Delta_{ij}, Z_{ij} \mid \theta_i + n^{-\frac{1}{2}} t_i, f, g_i),\ t = (t_1, ..., t_k) \in \mathbf{R}^k \tag{1.5}$$

and the $i$th component of $\gamma_n$ is given by

$$\gamma_{n,i} = -\frac{1}{n}\sum_{j=1}^{n_i}\{\Delta_{ij}\frac{f'}{f}(Z_{ij} - \theta_i) - (1 - \Delta_{ij})\frac{f}{F}(Z_{ij} - \theta_i)\}. \tag{1.6}$$

For details see Example 5.2 in Fabian and Hannan (1986). Thus an estimate $\hat{\theta}_n = (\hat{\theta}_{n,1}, ..., \hat{\theta}_{n,k})$ which satisfies

$$n^{\frac{1}{2}}(\hat{\theta}_n - \theta) - I^{-1}\gamma_n \to 0 \text{ in } P_{\theta,\eta} - \text{prob.}, \tag{1.7}$$

or equivalently,

$$n^{\frac{1}{2}}(\hat{\theta}_{n,i} - \theta_i) - I_i^{-1}\gamma_{n,i} \to 0 \text{ in } P_{\theta,\eta} - \text{prob., for each } i = 1, ..., k, \tag{1.8}$$

is locally asymptotically minimax at $\theta$ in the sense of Fabian and Hannan (1982, Def. 3.1) by their Theorem 6.3. Moreover, such an estimate is locally asymptotically minimax $\mathcal{A}$-adaptive at $(\theta, \eta)$ in the sense of Fabian and Hannan (1982, Def. 7.6) for each class $\mathcal{A}$ of LAN $(n\tilde{M}, \tilde{\gamma}_n)$ subproblems with $\tilde{M}_{12} = 0$ by their Theorem 7.10. Such a class $\mathcal{A}$ will be described below. Estimates satisfying (1.8) for a known $f$ and all $\theta$ and $g_1, ..., g_k$ subject to the conditions (A.3) and (A.4) have been constructed in Schick, Susarla and Koul (1986). This covers the case when $f$ is known.

We now consider the case when $f$ is unknown. It is easy to check that $\theta$ is not identifiable in this case. To circumvent this difficulty we require that for some known nondecreasing function $\psi$ from R to R satisfying $x\psi(x) > 0$ for all $x \neq 0$,

$$\int \psi f d\lambda = 0. \tag{A.5}$$

Choices of $\psi$ include $\psi(x) = x$ in which case $\theta_i$ is identified as the mean of $\Delta_{i1}Z_{i1}/G(Z_{i1})$ and $\psi(x) = -\frac{1}{q}[x < 0] + \frac{1}{1-q}[x > 0]$ for some $0 < q < 1$, in which case $\theta_i$ is identified as $q$th quantile of the survival function $F_i = P[Z_{i1} > \cdot]/G_i$. We also impose the following additional conditions on $f$ and $g_1, ..., g_k$:

$$\int \psi f' d\lambda \neq 0 \tag{A.6}$$

and

$$\max_{1\leq i\leq k} \int \frac{\psi^2 f}{G_i(\cdot + \theta_i + t)} d\lambda < \infty \quad \text{for a } t > 0. \tag{A.7}$$

We note that (A.5) to (A.7) imply the existence of $\sqrt{n}$-consistent estimates for $\theta$ under mild additional assumptions on $\psi$. For the two choices of $\psi$ given above the Kaplan-Meier

(1958) mean and qth quantile, respectively, based on the ith subsample are $\sqrt{n}$-consistent estimates of $\theta_i$. These preliminary estimates of $\theta_i$ are also asymptotically efficient for the ith subproblem as shown by Schick, Susarla and Koul (1986). However, we show here that the vector of estimates is generally not efficient for $\theta$.

The rest of this paper is organized as follows. In section 2 we address the question of how well one can estimate the parameter when $f$ is unknown. Following the lines of Schick (1986) we provide lower bounds on estimates of $\theta$ and give a sufficient condition for an estimate to obtain the lower bound. In section 3 we then treat the question of how to construct such estimates in the case when the censoring distributions are known.

## 2. *Efficiency Considerations*

We shall now investigate how well we can estimate $\theta$ asymptotically based on the data $\{(\Delta_{ij}, Z_{ij}) : j = 1, ..., n_i, i = 1, ..., k\}$ generated by $P_{\theta,\eta}$ if it is only known that the data are generated by a measure in $\mathcal{P} = \{P_{t,u} : t \in \mathbb{R}^k, u \in \mathcal{U}\}$ for a subset $\mathcal{U}$ of $\mathcal{F} \times \mathcal{G}^k$ with $\eta \in \mathcal{U}$. For example, if $f$ is known and $(g_1, ..., g_k)$ is unknown we can take $\mathcal{U} = \mathcal{U}_0 = \{f\} \times \mathcal{G}^k$; if $f$ is unknown and $(g_1, ..., g_k)$ is known we can take $\mathcal{U} = \mathcal{U}_1 = \mathcal{F}_\psi \times \{(g_1, ..., g_k)\}$ and if both $f$ and $(g_1, ..., g_k)$ are unknown we can take $\mathcal{U} = \mathcal{U}_2 = \mathcal{F}_\psi \times \mathcal{G}^k$, where $\mathcal{F}_\psi = \{\phi \in \mathcal{F} : \int \psi\phi d\lambda = 0\}$. We assume (A.1) - (A.7) throughout.

We begin by describing more precisely what we mean by asymptotically efficient estimates. By a ($q$-dimensional) ***path*** we mean a function $\beta$ on $\mathbb{R}^q$ into $\mathcal{U}$ such that $\beta(0) = \eta$. If $\beta$ is a path, we define functions $\beta_0, ..., \beta_k$ by $\beta(v) = (\beta_0(v), ..., \beta_k(v))$, $v \in \mathbb{R}^q$, and set

$$\Lambda_n(w \mid \beta) = \sum_{i=1}^{k}\sum_{j=1}^{n_i} \log p(\Delta_{ij}, Z_{ij} \mid \theta_i + n^{-\frac{1}{2}} w_i, \beta_0(n^{-\frac{1}{2}} v), \beta_i(n^{-\frac{1}{2}} v))$$

for $w = (w_1, ..., w_{k+q}) \in \mathbb{R}^{k+q}$ with $v = (w_{k+1}, ..., w_{k+q})$. We say a path is LAN if there are a positive definite $(k+q) \times (k+q)$ matrix $M(\beta)$ and $(k+q)$-dimensional random vectors $X_n(\beta)$ such that

$$\mathcal{L}(X_n(\beta) \mid P_{\theta,\eta}) \Longrightarrow \mathcal{N}(0, M(\beta))$$

and

$$\Lambda_n(w_n \mid \beta) - \Lambda_n(0 \mid \beta) - w_n^T X_n(\beta) - \frac{1}{2} w_n^T M(\beta) w_n \to 0 \text{ in } P_{\theta,\eta}\text{-prob.}$$

for every bounded sequence $w_n$ in $\mathbb{R}^{k+q}$. In this case, we partition the matrix $M(\beta)$ as

$$M(\beta) = \begin{bmatrix} M_{11}(\beta) & M_{12}(\beta) \\ M_{21}(\beta) & M_{22}(\beta) \end{bmatrix},$$

where $M_{11}(\beta)$ is $k \times k$, and set

$$V(\beta) = M_{11}(\beta) - M_{12}(\beta) M_{22}^{-1}(\beta) M_{21}(\beta). \tag{2.1}$$

It can be shown that $M_{11}(\beta) = I$. Let $\mathcal{B}$ be a family of LAN paths. A path $\beta^*$ is called ***least favorable in*** $\mathcal{B}$ if $\beta^* \in \mathcal{B}$ and if $V(\beta^*) \le V(\beta)$ for all $\beta \in \mathcal{B}$. (We write $A \le B$ for two square matrices $A, B$ of the same dimension if $B - A$ is nonnegative definite.) Suppose now $\mathcal{B}$ contains a least favorable path $\beta^*$. Then we call an estimate $T_n$ *$\mathcal{B}$-optimal* if, for every $\beta \in \mathcal{B}$,

$$\mathcal{L}(n^{\frac{1}{2}}(T_n - t_n) \mid P_{t_n,\beta(v_n)}) \Longrightarrow \mathcal{N}(0, V(\beta^*))$$

whenever $n^{\frac{1}{2}}(||t_n - \theta|| + ||v_n||)$ is bounded. A $\mathcal{B}$-optimal estimate is *locally asymptotically minimax in* $\mathcal{S}$ in the sense of Schick (1986b, Def. 4.12) by his Theorem 4.14, where $\mathcal{S}$ is the family of subproblems associated with $\mathcal{B}$.

We shall now exhibit families $\mathcal{B}$ of LAN paths which have a least favorable path $\beta^*$ for the above mentioned choices of $\mathcal{U}$ and then provide a characterization of $\mathcal{B}$-optimal estimates. For this we shall introduce additional notation. For $i = 1, ..., k$, let $Q_i = \lambda_i G(\cdot + \theta_i -)$, $K_i = \{h \in L_2(G_i) : \int h g_i d\mu = 0\}$ and $H_i = \{h \in L_2(p_i d\nu) : \int h p_i d\nu = 0\}$ and define, for $h = (h_1, ..., h_q) \in K_i^q$ a vector $r_i(\cdot \mid h) \in H_i^q$ by

$$r_i(\delta, z \mid h) = \delta \frac{1}{G_i(z)} \int_{(z,\infty]} h g_i d\mu + (1 - \delta) h(z), \ (\delta, z) \in S.$$

Set $Q = Q_1 + \ldots + Q_k$. For $t \in \mathbf{R}$ and $h = (h_1, \ldots, h_q) \in L_2^q(Qfd\lambda)$, define a function $l_t(\cdot \mid h)$ on $S$ into $\mathbf{R}^q$ by

$$l_t(\delta, z \mid h) = \delta h(z - t) + (1 - \delta) \frac{1}{F(z - t)} \int_{z-t}^{\infty} h f d\lambda, \ \ (\delta, z) \in S.$$

Set $\mathcal{H} = \{h \in L_2(Qfd\lambda) : l_{\theta_i}(\cdot \mid h) \in H_i, \ i = 1, \ldots, k\} = \{h \in L_2(Qfd\lambda) : \int hfd\lambda = 0, \ -\int (\int_{\cdot}^{\infty} hfd\lambda)^2 \frac{1}{F} dQ < \infty\}$. Verify that for $i = 1, ..., k$, $\xi \in \mathcal{H}$ and $\xi_i \in K_i$,

$$\int l_{\theta_i}(\cdot \mid \xi) r_i(\cdot \mid \xi_i) p_i d\nu = 0. \tag{2.2}$$

Let $K_0 = \{h \in \mathcal{H} : \int \psi h f d\lambda = 0\}$. For $h = (h_1, \ldots, h_q) \in K_0^q$, define a $k \times q$ matrix $C(h)$ by

$$C(h) = [\lambda_i \int l_{\theta_i}(\cdot \mid \frac{f'}{f}) l_{\theta_i}(\cdot \mid h_j) p_i d\nu].$$

We say a path is *smooth* if there are $\beta_i^\bullet \in K_i^q$, $i = 0, \ldots, k$ such that

$$\sum_{i=1}^{k} ||\beta_i^{1/2}(v) - \beta_i^{1/2}(0) - \frac{1}{2} g_i^{1/2} v^T \beta_i^\bullet||_\mu = o(||v||)$$

and

$$n \sum_{i=1}^{k} ||p^{\frac{1}{2}}(\cdot \mid \theta_i + t_{n,i}, \beta_0(v_n), g_i) - p_i^{\frac{1}{2}} - \frac{1}{2} p_i^{\frac{1}{2}} \{t_{n,i} l_{\theta_i}(\cdot \mid -\frac{f'}{f}) + v_n^T l_{\theta_i}(\cdot \mid \beta_0^\bullet)\}||_\nu^2 \to 0$$

whenever $n^{\frac{1}{2}}(||t_n|| + ||v_n||)$ is bounded.

We now list some properties of smooth paths.

(P.1) For each $\xi_i \in K_i^q$, $i = 0, \ldots, k$, there exists a smooth $\mathcal{U}_2$-valued path $\beta$ with

$\beta_i^\bullet = \xi_i$, $i = 0, \ldots, k$.

(P.2) Similar to Lemma 3.1 in Schick, Susarla and Koul (1986) one obtains, for a smooth path $\beta$, with $s_i(x,v) = p^{\frac{1}{2}}(\cdot \mid \theta_i + x, \beta_0(v), \beta_i(v))$,

$$n\sum_{i=1}^{k} ||s_i(t_{n,i}, v_n) - p_i^{\frac{1}{2}} - \frac{1}{2}p_i^{\frac{1}{2}}\{t_{n,i}l_{\theta_i}(\cdot \mid -\frac{f'}{f}) + v_n^T(l_{\theta_i}(\cdot \mid \beta_0^\bullet) + r_i(\cdot \mid \beta_i^\bullet))\}||_\nu^2 \to 0$$

whenever $n^{\frac{1}{2}}(||t_n|| + ||v_n||)$ is bounded. Thus by Theorem 4.5 in Fabian and Hannan (1986), a smooth path $\beta$ is LAN with

$$X_n(\beta) = (\gamma_n, \frac{1}{n}\sum_{i=1}^{k}\sum_{j=1}^{n_i}\{l_{\theta_i}(\Delta_{ij}, Z_{ij} \mid \beta_0^\bullet) + r_i(\Delta_{ij}, Z_{ij} \mid \beta_j^\bullet)\}$$

and

$$M(\beta) = \lim_n E_{\theta,\eta} X_n(\beta) X_n^T(\beta) = \begin{bmatrix} I & -C(\beta_0^\bullet) \\ -C^T(\beta_0^\bullet) & W(\beta) \end{bmatrix}$$

provided $M(\beta)$ is nonsingular, where

$$W(\beta) = \sum_{i=1}^{k} \int (l_{\theta_i}(\cdot \mid \beta_0^\bullet) + r_i(\cdot \mid \beta_i^\bullet))(l_{\theta_i}(\cdot \mid \beta_0^\bullet) + r_i(\cdot \mid \beta_i^\bullet)^T p_i d\nu$$

(P.3) For a smooth path $\beta$, the matrix $M(\beta)$ is nonsingular if and only if $W(\beta)$ is nonsingular.

For $a = 0, 1, 2$, let $\mathcal{B}_a$ be the family of all smooth $\mathcal{U}_a$-valued paths $\beta$ with $W(\beta)$ nonsingular. Then each $\mathcal{B}_a$ is a family of LAN paths. We shall now show that each $\mathcal{B}_a$ possesses a least favorable path.

We consider $\mathcal{B}_0$ first. For $\beta \in \mathcal{B}_0$, $\beta_0^\bullet = 0$ and thus $C(\beta_0^\bullet) = 0$. Consequently the necessary condition for adaptive estimation holds (see e.g. Condition 7.6 in Fabian and Hannan (1982)). Since $V(\beta) = I$ for each $\beta \in \mathcal{B}_0$, each path in $\mathcal{B}_0$ is least favorable. A $\mathcal{B}_0$-optimal estimate is described by (1.8).

Next, consider $\mathcal{B}_1$. To find a least favorable path in $\mathcal{B}_1$, we have to minimize $V(\beta)$ (see (2.1)) or equivalently maximize $C(\beta_0^\bullet)W^{-1}(\beta)C^T(\beta_0^\bullet)$. To carry out the details let $\nu_0$ denote the measure on S satisfying

$$\int h d\nu_0 = \int h(1, \cdot) Q f d\lambda - \int h(0, \cdot) f dQ$$

for nonnegative measurable function $h$ on $S$. Let

$$\Psi = \frac{1}{F}\int_{\cdot}^{\infty} \psi f d\lambda \tag{2.3}$$

and define vectors in $\mathcal{H}$ by

$$\psi^* = \frac{\psi}{Q} - \int_{-\infty}^{\cdot} \Psi d\frac{1}{Q} \tag{2.4}$$

and

$$\xi_i^* = \frac{Q_i}{Q}\frac{f'}{f} + \int_{-\infty}^{\cdot} \frac{f}{F} d\frac{Q_i}{Q}, \quad i = 1, \ldots, k. \tag{2.5}$$

Verify that

$$l_0(\delta, z \mid \psi^*) = \delta \frac{\psi}{Q}(z) + (1-\delta)\frac{\Psi}{Q}(z) - \int_{-\infty}^{z} \Psi d\frac{1}{Q} \tag{2.6}$$

and

$$l_0(\delta, z \mid \xi_i^*) = \frac{Q_i}{Q}(z)\left(\delta\frac{f'}{f}(z) - (1-\delta)\frac{f}{F}(z)\right) + \int_{-\infty}^{z} \frac{f}{F} d\frac{Q_i}{Q}, \quad i = 1, \ldots, k. \tag{2.7}$$

Let $\beta \in \mathcal{B}_1$. Then $\beta_i^\bullet = 0$, $i = 1, \ldots, k$ (since $\beta_i \equiv g_i$) and

$$W(\beta) = \sum_{i=1}^{k} \lambda_i \int l_{\theta_i}(\cdot \mid \beta_0^\bullet) l_{\theta_i}^T(\cdot \mid \beta_0^\bullet) p_i d\nu = \int l_0(\cdot \mid \beta_0^\bullet) l_0^T(\cdot \mid \beta_0^\bullet) d\nu_0.$$

Integration by parts yields

$$C(\beta_0^\bullet) = \int l_0(\cdot \mid \xi^*) l_0^T(\cdot \mid \beta_0^\bullet) d\nu_0$$

where $\xi^* = (\xi_1^*, \ldots, \xi_k^*)$. Thus, for $x \in \mathbf{R}^k$, $x^T C(\beta_0^\bullet) W^{-1}(\beta) C^T(\beta_0^\bullet) x$ denotes the squared $L_2(\nu_0)$-norm of the projection of $x^T l_0(\cdot \mid \xi^*) = l_0(\cdot \mid x^T \xi^*)$ onto the subset $T_\beta = \{v^T l_0(\cdot \mid \beta_0^\bullet) : v \in \mathbf{R}^q\}$ of $L_2(\nu_0)$. Let $T = \{l_0(\cdot \mid h) : h \in K_0\}$. The projection of $l_0(\cdot \mid h)$, $h \in \mathcal{H}$, onto $T$ exists and is given by $l_0(\cdot \mid h) - b(h) l(\cdot \mid \psi^*)$ where

$$b(h) = \int l_0(\cdot \mid h) l_0(\cdot \mid \psi^*) d\nu_0 \Big/ \int l_0^2(\cdot \mid \psi^*) d\nu_0.$$

This has been shown in Schick, Susarla and Koul (1986). Thus the projection of $l_0(\cdot \mid x^T \xi^*)$ onto $T$ is $l_0(\cdot \mid x^T(\xi^* - \psi^* \alpha))$, where

$$\alpha = (b(\xi_1^*), \ldots, b(\xi_k^*)) \tag{2.8}$$

and its squared $L_2(\nu_0)$-norm is $x^T J_* x$ where

$$J_* = \int l_0(\cdot \mid \xi^*) l_0^T(\cdot \mid \xi^*) d\nu_0 - \int l_0^2(\cdot \mid \psi^*) d\nu_0 \alpha \alpha^T. \tag{2.9}$$

This shows that, for each $x \in \mathbf{R}^k$,

$$x^T C(\beta_0^\bullet) W^{-1}(\beta) C^T(\beta_0^\bullet) x \leq x^T J_* x \tag{2.10}$$

with equality if $T_\beta$ contains the set $T_* = \{x^T l_0(\cdot \mid \xi^* - \psi^* \alpha) : x \in \mathbf{R}^q\}$. Consequently, every path $\beta \in \mathcal{B}_1$ which satisfies $T_\beta \supset T_*$ is least favorable. Of course, such a path exists by (P.1).

Now consider $\mathcal{B}_2$. Denote by $H_*$ the Hilbert space $H_1 \times \ldots \times H_k$ with inner product $\langle a,b\rangle_* = \sum_{i=1}^k \lambda_i \int a_i b_i p_i d\nu$. For $\xi \in \mathcal{H}$, let $l(\cdot \mid \xi) = (l_{\theta_1}(\cdot \mid \xi), \ldots, l_{\theta_k}(\cdot \mid \xi)) \in H_*$, and, for $x \in \mathbf{R}^q$, let $\varsigma(x) = (x_1 l_{\theta_1}(\cdot \mid \frac{f'}{f}), \ldots, x_k l_{\theta_k}(\cdot \mid \frac{f'}{f})) \in H_*$. Define subsets of $H_*$ by $D_0 = \{l(\cdot \mid \xi) : \xi \in K_0\}$ and $D_1 = \{(r_1(\cdot \mid \xi_1), \ldots, r_k(\cdot \mid \xi_k)) : \xi_i \in K_i,\ i = 1, \ldots, k\}$. In view of (2.2), $D_0 \cup \{\varsigma(x) : x \in \mathbf{R}^k\}$ and $D_1$ are orthogonal subsets of $H_*$. Moreover, for $\xi_1$ and $\xi_2$ in $K_0$,

$$\langle l(\cdot \mid \xi_1), l(\cdot \mid \xi_2)\rangle_* = \int l_0(\cdot \mid \xi_1) l_0(\cdot \mid \xi_2) d\nu_0$$

and

$$\langle l(\cdot \mid \xi_1), \varsigma(x)\rangle_* = \int l_0(\cdot \mid \xi_1) l_0(\cdot \mid x^T \xi^*) d\nu_0.$$

Thus the projection of $\varsigma(x)$ onto $D = D_0 + D_1$ equals the projection of $\varsigma(x)$ onto $D_0$ and is given by $l(\cdot \mid x^T(\xi^* - \psi^*\alpha))$. Now let $\beta \in \mathcal{B}_2$. Set $s_i = l_{\theta_i}(\cdot \mid \beta_0^\bullet) + r_i(\cdot \mid \beta_i^\bullet)$, $i = 1, \ldots, k$ and $D_\beta = \{(v^T s_1, \ldots, v^T s_k) : v \in \mathbf{R}\}$. Then $x^T C(\beta_0^\bullet) W^{-1}(\beta) C^T(\beta_0^\bullet) x$ is the squared norm of the projection of $\varsigma(x)$ onto $D_\beta \subset H_*$ and is less than or equal to the squared norm $x^T J_* x$ of the projection of $\varsigma(x)$ onto $D$. Thus (2.10) holds, for $x \in \mathbf{R}^k$, with equality if $D_\beta$ contains the set $D_* = \{l(\cdot \mid x^T(\xi^* - \psi^*\alpha)) : x \in \mathbf{R}^k\}$. Consequently, every path $\beta \in \mathcal{B}_2$ which satisfies $D_\beta \supset D_*$ is least favorable.

The above show that a path $\beta$ is least favorable in $\mathcal{B}_1$ or $\mathcal{B}_2$ if $V(\beta) = I - J_*$. Consequently, a path that is least favorable in $\mathcal{B}_1$ is also least favorable in $\mathcal{B}_2$. Note that $J_*$ measures the loss of information for not knowing $f$. There is no loss for not knowing $(g_1, \ldots, g_k)$. See also Ritov (1985).

We shall now characterize $\mathcal{B}_1$- and $\mathcal{B}_2$-optimal estimates. For this, let

$$Y_n = \theta + (n(I - J_*))^{-1} \sum_{i=1}^{k} \sum_{j=1}^{n_i} l_{\theta_i}(\Delta_{ij}, Z_{ij} \mid h_i^*)$$

where

$$h_i^* = -\frac{f'}{f} e_i + \xi^* - \psi^* \alpha$$

and $e_i$ is the $i$th standard unit vector in $\mathbf{R}^k$.

**Theorem 2.1.** Let $T_n$ be an estimate. Then the following statements are equivalent.
(a) $T_n$ is $\mathcal{B}_1$-optimal.
(b) $T_n$ is $\mathcal{B}_2$-optimal.
(c) $n^{\frac{1}{2}}(T_n - Y_n) \to 0$ in $P_{\theta,\eta}$-prob..

**Proof.** That (c) implies (b) can be derived either from Theorem 4.19 in Schick (1986b) or by a direct argument using LeCam's Third Lemma (see Hájek and Šidák (1967)). Of course, (b) implies (a). That (a) implies (c) can be verified by a similar argument as in Bickel (1982, Theorem 6.1).

We shall now modify the construction in Schick (1986a) to apply to our present situation. We have to construct an estimate $T_n$ satisfying (c). Define quantities $Q_{n,t}$, $\xi_{n,t}^*$, $\psi_{n,t}^*$,

$\nu_{n,t}$, $\alpha_{n,t}$ and $V_n(t)$ by replacing in the definitions of $Q$, $\xi^*$, $\psi^*$, $\nu_0$, $\alpha$, and $I - J_*$, respectively, $\lambda_i$ by $n_i/n$ and $\theta$ by $t$. Let

$$h^*_{n,t,i} = -\frac{f'}{f}e_i + \xi^*_{n,t} - \psi^*_{n,t}\alpha_{n,t}$$

and

$$Z_n(t) = t + (nV_n(t))^{-1}\sum_{i=1}^{k}\sum_{j=1}^{n_i} l_{t_i}(\Delta_{ij}, Z_{ij} \mid h^*_{n,t,i})$$

The following conditions are analogous to (A.1) to (A.3) of Schick (1986a). We say $t_n$ is a *local* sequence if $t_n$ is a sequence in $\mathbb{R}^k$ such that $n^{\frac{1}{2}}(t_n - \theta)$ is bounded.

(C.1) For every local sequence $t_n$, (c) holds with $T_n = Z_n(t_n)$.
(C.2) $\hat{\theta}_n$ is a $n^{\frac{1}{2}}$-consistent estimate of $\theta$, i.e. $n^{\frac{1}{2}}(\hat{\theta}_n - \theta)$ is bounded in $P_{\theta,\eta}$-prob..
(C.3) There are measurable functions $L_{n,i}$ on $S \times \mathbb{R}^k \times S^{n_1} \times \ldots \times S^{n_k}$ to $\mathbb{R}^k$ such that for every local sequence $t_n = (t_{n1}, \ldots, t_{nk})$

$$n^{-\frac{1}{2}}\sum_{i=1}^{k} n_i \int \hat{L}_{n,i}(\cdot, t_n)p(\cdot \mid t_{ni}, f, g_i)d\nu \to 0 \text{ in } P_{t_n,\eta}\text{-prob.} \tag{2.11}$$

and

$$\sum_{i=1}^{k}\int ||\hat{L}_{n,i}(\cdot, t_n) - l_{t_{ni}}(\cdot \mid h^*_{n,t_n,i})||^2\, p(\cdot \mid t_{ni}, f, g_i)\, d\nu \to 0 \text{ in } P_{t_n,\eta}\text{-prob.} \tag{2.12}$$

with $\hat{L}_{n,i}(\cdot,\cdot) = L_{n,i}(\cdot,\cdot,S_{n1},\ldots,S_{nk})$ and $S_{ni} = ((\Delta_{i1}, Z_{i1}), \ldots, (\Delta_{in_i}, Z_{in_i}))$.

Under these conditions an estimate satisfying (c) can be constructed as follows. Discretize $\hat{\theta}_n$ and denote the discretized version by $\bar{\theta}_n$. Split each subsample into two parts. Let $S^{(1)}_{n,i} = ((\Delta_{i1}, Z_{i1}), \ldots, (\Delta_{im_i}, Z_{im_i}))$ and $S^{(2)}_{ni} = ((\Delta_{im_i+1}, Z_{im_i+1}), \ldots, (\Delta_{in_i}, Z_{in_i}))$ denote the parts of the $i$th subsample and set $X^{(1)}_{nij} = L_{m,i}((\Delta_{ij}, Z_{ij}), \bar{\theta}_n, S^{(1)}_{n1}, \ldots, S^{(1)}_{nk})$, and $X^{(2)}_{nij} = L_{n-m,i}((\Delta_{ij}, Z_{ij}), \bar{\theta}_n, S^{(2)}_{n1}, \ldots, S^{(2)}_{nk})$.

**Theorem 2.2.** Suppose (C.1) to (C.3) hold and $m_i/n_i \to \frac{1}{2}$, $i = 1, \ldots, k$. Then the estimate

$$\bar{\theta}_n + \hat{V}_n^- \sum_{i=1}^{k}\Big(\sum_{j=1}^{m_i} X^{(2)}_{nij} + \sum_{j=m_i+1}^{n_i} X^{(1)}_{nij}\Big)$$

satisfies (c) of Theorem 2.1 where $\hat{V}_n^-$ denotes a generalized inverse of

$$\hat{V}_n = \sum_{i=1}^{k}\Big\{\sum_{j=1}^{m_i} X^{(2)}_{nij}(X^{(2)}_{nij})^T + \sum_{j=m_i+1}^{n_i} X^{(1)}_{nij}(X^{(1)}_{nij})^T\Big\}.$$

We conclude this section with a few remarks. Condition (C.1) can be verified if $G_1, \ldots, G_k$ are continuous and $\psi$ is smooth. Estimates satisfying (C.2) have been exhibited in Schick, Susarla, and Koul (1986) under smoothness conditions on $\psi$. In the next section, we provide functions $L_{n,i}$ as required in (C.3) in the case when $g_1, \ldots, g_k$ are known. In this case, Condition (2.11) can be achieved easily. If $g_1, \ldots, g_k$ are unknown, then the construction of $L_{n,i}$ is considerably more difficult mainly because of verification of (2.11).

## §3. *Construction of efficient estimates when $g_1, \ldots, g_k$ are known.*

We shall now construct functions $L_{n,i}$ as described in (C.3) above. Let $F_{n,i}$ denote the product-limit estimator based on $\{(\Delta_{ij}, Z_{ij}),\ j = 1, \ldots, n_i\}$ for $i = 1, \ldots, k$ and set $\tilde{F}_{n,t} = \sum_{i=1}^{k}(n_i/n)F_{n,i}(\cdot + t_i)$ for $t = (t_1, \ldots, t_k) \in \mathbf{R}^k$. Using the results of Gill (1983) together with (A.7), we can show that for each local sequence $t_n$

$$\|\tilde{F}_{n,t_n} - F\|_\infty = O_{t_n}(n^{-1/2}). \tag{3.1}$$

Throughout we let $o_{t_n}(1)$ and $O_{t_n}(1)$ denote convergence to zero and boundedness in $P_{t_n,\eta}$-probability.

(3.2) Let $a_n$, $b_n$, and $c_n$ be positive numbers going to zero such that $na_n^6 \to \infty$, $na_n^2 b_n^4 \to \infty$, and $nc_n^2 \to \infty$.

Let $K$ be the logistic density, i.e. $K(s) = e^{-s}/(1+e^{-s})^2 = e^s/(1+e^s)^2$, $-\infty < s < \infty$ and $K_n(\cdot) = a_n^{-1} K(\cdot/a_n)$. For $t \in \mathbf{R}^k$ and $z \in \mathbf{R}$, set

$$A_{n,t}(z) = \frac{1}{Q_{n,t}(z)}\left(\frac{n_1}{n}G_1(z+t_1-), \ldots, \frac{n_k}{n}G_k(z+t_k-)\right) \tag{3.3}$$

$$\hat{f}_{n,t}^{(r)}(z) = -\int K_n^{(r)}(z-\cdot)\,d\tilde{F}_{n,t}, \quad r = 0, 1 \tag{3.4}$$

$$\hat{F}_{n,t}(z) = \int_z^\infty \hat{f}_{n,t}\,d\lambda, \tag{3.5}$$

$$D_{n,t}(z) = \frac{1}{n}\sum_{i=1}^{k}\sum_{j=1}^{n_i} \Delta_{ij}\frac{\psi}{Q_{n,t}}(Z_{ij} - t_i)[Z_{ij} - t_i > z], \tag{3.6}$$

and

$$\tilde{\Psi}_{n,t}(z) = \frac{D_{n,t}(z)}{\tilde{F}_{n,t}(z) + c_n}. \tag{3.7}$$

Now set

$$\rho_{n,t,i}(\delta, z) = -(e_i - A_{n,t}(z))\left\{\delta\frac{\hat{f}'_{n,t}(z)}{\hat{f}_{n,t}(z) + a_n} - (1-\delta)\frac{\hat{f}_{n,t}(z)}{\hat{F}_{n,t}(z) + b_n}\right\} + \int_{-\infty}^{z} \frac{\hat{f}_{n,t}}{\hat{F}_{n,t} + b_n}\,dA_{n,t}$$

and

$$\sigma_{n,t}(\delta, z) = \frac{1}{Q_{n,t}(z)}\{\delta\psi(z) + (1-\delta)\tilde{\Psi}_{n,t}(z)\} - \int_{-\infty}^{z} \tilde{\Psi}_{n,t} d\frac{1}{Q_{n,t}}$$

and define the function $L_{n,i}$ by

$$L_{n,i}((\delta, z), t, S_{n1}, \ldots, S_{nk}) = \rho_{n,t,i}(\delta, z - t_i) + \hat{\alpha}_{n,t}\sigma_{n,t}(\delta, z - t_i) \tag{3.8}$$

where $\hat{\alpha}_{n,t}$ is an estimate of $\alpha_{n,t}$. One verifies easily that for all $t$ in $\mathbf{R}^k$,

$$\sum_{i=1}^{k} n_i \int L_{n,i}(\cdot, t, S_{n1}, \ldots, S_{nk}) p(\cdot \mid t_i, f, g_i) d\nu = 0. \tag{3.9}$$

Consequently, (2.11) holds. (2.12) is implied if the following conditions hold for every local sequence $t_n = (t_{n1}, \ldots, t_{nk})$:

$$\hat{\alpha}_{n,t_n} - \alpha_{n,t_n} = o_{t_n}(1) \tag{3.10}$$

$$\int \left(\frac{\hat{f}'_{n,t_n}}{\hat{f}_{n,t_n} + a_n} - \frac{f'}{f}\right)^2 f d\lambda = o_{t_n}(1) \tag{3.11}$$

$$\int \left(\frac{\hat{f}_{n,t_n}}{\hat{F}_{n,t_n} + b_n} - \frac{f}{F}\right)^2 F dQ_{n,t_n} = o_{t_n}(1) \tag{3.12}$$

$$\max_{1 \le i \le k} \int \left[\int_{-\infty}^{\cdot} \left(\frac{\hat{f}_{n,t_n}}{\hat{F}_{n,t_n} + b_n} - \frac{f}{F}\right) d\frac{G_i(\cdot + t_{ni}-)}{Q_{n,t_n}}\right]^2 dFQ_{n,t_n} = o_{t_n}(1). \tag{3.13}$$

$$\int (\tilde{\Psi}_{n,t_n} - \Psi)^2 \; F d\frac{1}{Q_{n,t_n}} = o_{t_n}(1) \tag{3.14}$$

and

$$\int \left(\int_{-\infty}^{\cdot} \tilde{\Psi}_{n,t} - \Psi \; d\frac{1}{Q_{n,t_n}}\right)^2 \; dFQ_{n,t_n} = o_{t_n}(1). \tag{3.15}$$

Verification of these conditions requires analysis of $\hat{f}^{(i)}_{n,t}$, $\hat{F}_{n,t}$, and the following two functions:

$$f_n = \int K_n(\cdot - s) dF(s) \tag{3.16}$$

and

$$F_n = \int^{\infty} f_n d\lambda. \tag{3.17}$$

The first lemma deals with some properties of $f_n$ and $F_n$ which are of independent interest.

**Lemma 3.1.**

(a)
$$J(f_n) = \int f_n'^2/f_n \, d\lambda \le J(f) = \int f'^2/f \, d\lambda < \infty.$$

(b)
$$\|f_n'/f_n\|_\infty \le 1/a_n, \quad \|f_n/F_n\|_\infty \le 1/a_n,$$

(c)
$$f^2/F \le \int_\cdot^\infty \frac{f'^2}{f} d\lambda, \quad f_n^2/F_n \le \int_\cdot^\infty f_n'^2/f_n \lambda.$$

(d)
$$\int \left(\frac{f_n'}{f_n^{1/2}} - \frac{f'}{f^{1/2}}\right)^2 d\lambda \to 0.$$

(e)
$$a_n^{-2} \int (f_n^{1/2} - f^{1/2})^2 \, d\lambda \to 0.$$

(f)
$$\int \left(\frac{f_n'}{f_n} - \frac{f'}{f}\right)^2 f d\lambda \to 0.$$

(g)
$$|F_n^{1/2} - F^{1/2}|^2 \le 2 \int_\cdot^\infty (f_n^{1/2} - f^{1/2})^2 \, d\lambda.$$

(h)
$$\left\|\frac{f_n}{F_n^{1/2}} - \frac{f}{F^{1/2}}\right\|_\infty \to 0.$$

(i)
$$\left\|F^{1/2}\left(\frac{f_n}{F_n} - \frac{f}{F}\right)\right\|_\infty \to 0.$$

*Proof.* (a) can be found in Hájek and Šidák (1986, p. 17), (d) can be found in Bickel (1982), while (b) and (c) are trivial. We now prove (e) which is a stronger conclusion than given in Bickel (1982, (6.21) and (6.22)). To prove this result, let

$$v_a(\cdot) = \frac{1}{2} \int_0^1 \frac{\int K(u) \, f'(\cdot - aus) u d\lambda(u)}{[\int K(u) f(\cdot - aus) u d\lambda(u)]^{\frac{1}{2}}} d\lambda(s)$$

and

$$w_a(\cdot) = \frac{1}{4} \int_0^1 \left\{ \int_{-\infty}^\infty \frac{u f'(\cdot - aus)}{f(\cdot - aus)} K(u) d\lambda(u) \right\}^2 d\lambda(s).$$

Note that $v_{a_n} = a_n^{-1}(f_n^{1/2} - f^{1/2})$. Now, Bickel (1982, Proof of Lemma 6.3) shows that $v_a^2 \le w_a$ and that $\int w_a d\lambda \le (\frac{1}{4})J(f)\int u^2 K(u)d\lambda(u) = \int w_0 d\lambda$ for all $a \ge 0$. Moreover, as $a \to 0$, $w_a \to w_0$ and $v_a \to 0$, both almost everywhere $\lambda$. Therefore, an application of the extended version of the Lebesgue dominated convergence theorem gives the desired result.

(f) follows from (b), (d), (e) and

$$f(\frac{f_n'}{f_n} - \frac{f'}{f})^2 \le 2\{(\frac{f_n'}{f_n})^2(f_n^{1/2} - f^{1/2})^2 + (\frac{f_n'}{f_n^{1/2}} - \frac{f'}{f^{1/2}})^2\}.$$

(g) follows by applying the Cauchy Schwarz inequality to

$$F_n^{1/2} - F^{1/2} = \int_{\cdot}^{\infty} (f_n^{1/2} - f^{1/2})(f_n^{1/2} + f^{1/2})d\lambda/(F_n^{1/2} + F^{1/2}).$$

To prove (h), let $\epsilon > 0$ be arbitrary. Let $T_\epsilon$ be such that $\int_{T_\epsilon}^{\infty}(f'^2/f)d\lambda < \epsilon$. Now on $(-\infty, T_\epsilon]$, $f_n/F_n^{\frac{1}{2}}$ converges uniformly to $f/F^{\frac{1}{2}}$ while for $t > T_\epsilon$, by the last part of (c),

$$\frac{f_n^2}{F_n}(t) \le \int_t^{\infty} (\frac{f_n'^2}{f_n} - \frac{f'^2}{f})d\lambda + \epsilon.$$

The first term goes to zero uniformly in $t$ by (d). Since $\epsilon$ is arbitrary, the proof is complete.

(i) follows from (g), (h) and

$$|F^{1/2}(\frac{f_n}{F_n} - \frac{f}{F})| \le \frac{f_n}{F_n}|F_n^{1/2} - F^{1/2}| + |\frac{f_n}{F_n^{1/2}} - \frac{f}{F^{1/2}}|. \ \square$$

Before we state and prove the next lemma, we point out the following results. For each local sequence $t_n$, we obtain from (3.1) and integration by parts

$$||\hat{F}_{n,t_n} - F_n||_\infty = o_{t_n}(n^{-1/2}\, a_n^{-1}) \tag{3.18}$$

and

$$||\hat{f}_{n,t_n}^{(i)} - f_n^{(i)}||_\infty = O_{t_n}(n^{-1/2}\, a_n^{-(i+1)}) \tag{3.19}$$

With the above inequalities, we are now ready to prove

**Lemma 3.2.**

$(a)$ $\quad na_n^6 \to \infty$ implies (3.11)

$(b)$ $\quad na_n^2 b_n^4 \to \infty$ implies (3.12)

$(c)$ $\quad na_n^2 b_n^4 \to \infty$ implies (3.13)

*Proof.* Since the proofs (a) and (b) follow directly from Lemma 3.1, (3.18) and (3.19), we prove (c) only. Fix a local sequence $t_n$. Set

$$\hat{\gamma}_n = |\frac{\hat{f}_{n,t_n}}{\hat{F}_{n,t_n} + b_n} - \frac{f}{F}| \text{ and } Q_n = Q_{n,t_n}.$$

By the definition of $A_{n,t}$ it is enough to show that

$$T = \int (\int_{-\infty}^{\cdot} \frac{\hat{\gamma}_n}{Q_n} dQ_n)^2 dFQ_n = o_{t_n}(1). \tag{3.20}$$

Then we obtain for a constant $c > 0$

$$T \leq c \int\int_{s<u} F^{1/2}(s)\hat{\gamma}_n(s)\ F^{1/2}(u)\hat{\gamma}_n(u)\ dQ_n^{1/2}(s)dQ_n^{1/2}(u)$$

by writing the inside integral (3.20) as a double integral and then integrating w.r.t. $FQ_n$. Hence the result follows if $||F\hat{\gamma}_n^2||_\infty = o_{t_n}(1)$. Observe that

$$|F\hat{\gamma}_n^2| \leq 3\{[\frac{\hat{f}_{n,t_n}}{\hat{F}_{n,t_n} + b_n} - \frac{f_n}{F_n + b_n}]^2 + F(\frac{f_n}{F_n + b_n} - \frac{f_n}{F_n})^2 + F(\frac{f_n}{F_n} - \frac{f}{F})^2\} = 3(T_1^2 + T_2 + T_3).$$

By (i) of Lemma 3.1, $||T_3||_\infty \to 0$. By (3.18) and (3.19), $||T_1||_\infty = O_{t_n}(1/(n^{\frac{1}{2}}\ a_n b_n^2))$. That $||T_2||_\infty \to 0$ can be seen by starting with the inequality

$$F(\frac{f_n}{F_n + b_n} - \frac{f_n}{F_n})^2 \leq F(\frac{f_n}{F_n})^2(\frac{b_n}{F_n + b_n})^2 \leq 2\{T_3 + (\frac{f^2}{F})[b_n/(F_n + b_n)]^2\}$$

and observing that the second term on the right hand side goes to zero uniformly by (e) and (g) of Lemma 3.1. □

**Lemma 3.3.** $nc_n^2 \to \infty$ implies (3.14) and (3.15).

*Proof.* Fix a local sequence $t_n$. Abbreviate $Q_{n,t_n}$ by $Q_n$ and $\tilde{\Psi}_{n,t_n}$ by $\tilde{\Psi}_n$. By (A.7),

$$\int \frac{\psi^2}{Q_n} f d\lambda = O(1) \text{ and } \int F d\frac{1}{Q_n} \leq \int \frac{f}{Q_n} d\lambda = O(1) \tag{3.21}$$

Let $B_n = D_{n,t_n} - \int_{\cdot}^{\infty} \psi f d\lambda$. Then

$$E_{t_n,\eta} B_n = 0 \text{ and } E_{t_n,\eta} B_n^2 = \frac{1}{n}\int_{\cdot}^{\infty} \frac{\psi^2}{Q_n} f d\lambda.$$

This and (3.21) show that

$$E_{t_n,\eta} \int B_n^2 F d\frac{1}{Q_n} = O(\frac{1}{n}) \tag{3.22}$$

and

$$(3.23)\qquad \begin{aligned} -\frac{1}{2}E_{t_n,\eta}\int(\int_{-\infty}^{\cdot}|B_n|d\frac{1}{Q_n})^2 dFQ_n &\le E_{t_n,\eta}\int FQ_n|B_n|(\int_{-\infty}^{\cdot}|B_n|d\frac{1}{Q_n})d\frac{1}{Q_n} \\ &= O(\frac{1}{n}\int FQ_n(\int_{-\infty}^{\cdot}d\frac{1}{Q_n})d\frac{1}{Q_n}) = O(\frac{1}{n}) \end{aligned}$$

With $R_n = c_n/(F+c_n)$ or $1$, we have

$$(3.24)\qquad \int \Psi^2 R_n^2 F d\frac{1}{Q_n} \le \int(\int_{\cdot}^{\infty} R_n^2\psi^2 f d\lambda)d\frac{1}{Q_n} \le \int \frac{1}{Q_n}(R_n\psi)^2 f d\lambda$$

and, since for $0 < s < t$, $0 < \Psi(s) < \Psi(t) + \psi(t)$, and $\psi$ and $R_n$ are nondecreasing,

$$(3.25)\qquad \begin{aligned} -\frac{1}{2}\int_0^{\infty}(\int_0^{\infty}|\Psi|R_n d\frac{1}{Q_n})^2 dFQ_n &\le \int_0^{\infty} FQ_n|\Psi|R_n(\int_0^{\cdot}|\Psi|R_n d\frac{1}{Q_n})d\frac{1}{Q_n} \\ &\le \int_0^{\infty}\left(F\Psi^2R_n^2 + \int_{\cdot}^{\infty}(\psi R_n)^2 f d\lambda\right)d\frac{1}{Q_n} \\ &\le 2\int_0^{\infty}\frac{(\psi R_n)^2}{Q_n} f d\lambda. \end{aligned}$$

From (A.7) and the dominated convergence theorem we obtain

$$(3.26)\qquad \int \frac{\psi^2}{Q_n}(\frac{c_n}{F+c_n})^2 f d\lambda \to 0.$$

Now let

$$\Psi_n = \frac{\int_{\cdot}^{\infty}\psi f d\lambda}{F+c_n} = \Psi\frac{F}{F+c_n}.$$

Then

$$(3.27)\qquad c_n|\tilde{\Psi}_n - \Psi_n| \le |B_n| + ||\tilde{F}_{n,t_n} - F||_{\infty}|\Psi|.$$

Now, (3.24), (3.25), and (3.26) imply

$$\int(\Psi_n - \Psi)^2 F d\frac{1}{Q_n} \to 0 \text{ and } \int(\int_{-\infty}^{\cdot}|\Psi_n - \Psi|d\frac{1}{Q_n})^2 dFQ_n \to 0.$$

Next, (3.27), (3.22), (3.1), and (3.24) with $R_n = 1$ give

$$\int(\tilde{\Psi}_n - \Psi_n)^2 F d\frac{1}{Q_n} = O_{t_n}(n^{-1}c_n^{-2}) = o_{t_n}(1)$$

while (3.27), (3.23), (3.1), and (3.25) with $R_n = 1$ give

$$\int(\int_{-\infty}^{\cdot}|\tilde{\Psi}_n - \Psi_n|d\frac{1}{Q_n})^2 dFQ_n = O_{t_n}(n^{-1}c_n^{-2}) = o_{t_n}(1).$$

This proves (3.14) and (3.15). □

Estimates $\hat{\alpha}_{n,t}$ satisfying (3.10) can be obtained as follows. Split the sample as before. Use one part of the sample to estimate $l_0(\cdot \mid \xi^*_{n,t})$ in the same way as in $\rho_{n,t}$ and $l_0(\cdot \mid \Psi^*_{n,t})$ as in $\sigma_{n,t}$ and then mimic the integrals in the definition of $\alpha_{n,t}$ with the appropriate averages based on the remaining observations.

*Acknowledgement.* The research work of V. Susarla was supported in part by the National Institute of Health under grant No. 1R01GM28405.

## References

1. Begun, Janet M., Hall, W.J., Huang, Wei-Min, and Wellner, Jon A. (1983). 'Information and asymptotic efficiency in parametric-nonparametric models.' *Ann. Statist.* **11** 432-452.

2. Bickel, P.J. (1982). 'On adaptive estimation.' *Ann. Statist.* **10** 647-671.

3. Fabian, V. and Hannan, J. (1982). 'On estimation and adaptive estimation for locally asymptotically normal families.' *Z. Wahrsch. verw. Gebiete.* **59** 459-478.

4. ———— (1986). 'Local asymptotic behavior of densities.' To appear.

5. Gill, Richard (1983). 'Large sample behaviour of the product-limit estimator on the whole line.' *Ann. Statist.* **11** 49-58.

6. Hájek, J. and Šidák, Z. (1967). 'Theory of rank tests'. *Academic Press* New York and Academia, Prague.

7. Kaplan, E.L. and Meier, P. (1958). 'Nonparametric estimation from incomplete observations.' *J. Amer. Stat. Assoc.* **53** 457-481.

8. Ritov, Y. (1985). 'Efficient and unbiased estimation in nonparametric linear regression with censored data.' Tech. Rep. Dept. of Statistics. Univ. of Jerusalem.

9. Schick, A. (1986a). 'On asymptotically efficient estimation in semiparametric models.' *Ann. Statist.* **14**, 1139-1151.

10. ———— (1986b). 'On estimation in LAMN families when there are nuisance parameters present.' Submitted for publication.

11. Schick, A., Susarla, V. and Koul, H. (1986). 'Efficient estimation of location with censored data.' Submitted for publication.

# AN OPTIMAL k-STOPPING PROBLEM FOR THE POISSON PROCESS

Wolfgang Stadje
Fachbereich Mathematik/Informatik
Universität Osnabrück
Albrechtstr. 28
4500 Osnabrück
West Germany

ABSTRACT. Offers for k commodities of the same kind arrive according to a Poisson process. It is assumed that these offers are i.i.d. random variables discounted by some deterministic function of time. Generalizing the well-known case k=1 an optimal strategy for the seller is found. His optimal k-stopping rule is studied in detail and explicitly computed in some examples.

## I. INTRODUCTION

The owner of k commodities of the same kind wants to sell them. Sequentially he receives offers which he has to refuse or accept immediately on arrival. In what follows, a probabilistic model for this situation is presented, and an optimal strategy for the seller is found and explicitly computed in some examples.

We assume that the offers are of random size and arrive at random time points. Thus we are confronted with a version of the so-called (multiple) best choice problems which date back to the now classic secretary problem (see the review of Freeman (1983) containing a (up to then) complete list of references). However we do not consider the question of maximizing some probability (for example that of getting the k best offers); the problem treated here is to maximize the discounted expected gain of the seller. The case of only one commodity for sale is treated in the monograph of Chow, Robbins and Siegmund (1971, pages 113-118). The solution was first given by Elfving (1967) and later justified by Siegmund (1967) who removed some shortcomings. We generalize their model to the case $k > 1$. An interesting extension of the best choice problem with discrete time and complete knowledge of the distribution of the offers is Petruccelli (1982). Various aspects of selecting sequentially k out of n independent random values coming from a known distribution have been treated by Henke (1970a, 1970b), Tamaki (1979), Móri (1983, 1984), Stadje (1985), and, in the context of prophet inequalities, by Kennedy (1986). Among the references to optimal stopping problems associated to a Poisson process we mention Cowan and Zabczyk (1978), Cieselski and Zabczyk (1979), Sakaguchi (1976), Gaver (1976),



*P. Bauer et al. (eds.), Mathematical Statistics and Probability Theory, Vol. B, 231–244.*

and Bruss (1986).

## II. PROPERTIES OF THE OPTIMAL k-STOPPING RULE

The model we shall now present coincides with that of Chow et al. (1971). Let $Y_1, Y_2, \ldots$ be independent, non-negative random variables with common piecewise continuous distribution function (d.f.) $F(x)$ and mean $\mu \in (0,\infty)$. Let $0<\tau_1<\tau_2\ldots$ be the time points of jump of a homogeneous Poisson process with intensity $\lambda > 0$. $\tau_n$ represents the time of the n-th offer and $Y_n$ the offered amount. The sequences $(Y_n)_{n\geq 1}$ and $(\tau_n)_{n\geq 1}$ are assumed to be independent. We remark that by a change of scale the case of an inhomogeneous Poisson process can be reduced to our problem. Further there is given a discount function $r: [0,\infty)\to[0,1]$ with the following properties: r is non-increasing, continuous from the right, piecewise continuous, and satisfies the relations

$$r(0) = 1 \tag{1}$$

$$\int_0^\infty r(u)\,du < \infty \tag{1'}$$

$$r(u+v) \leq r(u)r(v) \quad \text{for all } u,v \geq 0. \tag{1"}$$

Inequality (1") e.g. holds, if r is of the form $r(u) = e^{-f(u)} 1_{[0,T)}(u)$, where $f \geq 0$ satisfies $f(u) + f(v) \leq f(u+v)$ and $0 < T \leq \infty$. Typical examples are $r(u) = e^{-\alpha u}$ or $r = 1_{[0,T)}$, $T < \infty$. If the seller accepts the offers no. $n_1,\ldots,n_k$, he receives an amount of

$$X_{n_1,\ldots,n_k} := \sum_{i=1}^{k} Y_{n_i} r(\tau_{n_i}).$$

His problem is to choose a "k-stopping rule", i.e. k stopping times $t_1 < t_2 < \ldots < t_k$ relative to the sequence $\mathcal{F}_n := \sigma(Y_1,\ldots,Y_n,\tau_1,\ldots,\tau_n)$ of $\sigma$-algebras such that $E(X_{t_1,\ldots,t_k})$ is maximal among all such k-stopping rules.

There is a natural candidate for an optimal k-stopping rule. Suppose that already the j-1 offers $n_1,\ldots,n_{j-1}$ are accepted and the seller now has to decide whether he refuses the n-th offer or not. Then he should accept iff

$$\text{ess sup}\{E(X_{n_1,\ldots,n_{j-1},n,t_{j+1},\ldots,t_k}|F_n)|n<t_{j+1}<\ldots<t_k\}$$

$$\geq \text{ess sup}\{E(X_{n_1,\ldots,n_{j-1},t_j,\ldots,t_k}|F_n)|n<t_j<\ldots<t_k\},$$

the lefthand side being its maximal conditional expected reward, if he takes the n-th offer, whereas the righthand side is the analogous reward for the case that he refuses the n-th offer. Conditions for this strategy to be optimal are given in Stadje (1985). Note that the left-hand side is equal to

$$\sum_{i=1}^{j-1} Y_{n_i} r(\tau_{n_i}) + Y_n r(\tau_n) + \text{ess sup}\{E(\sum_{i=1}^{k-j} Y_{t_i} r(t_i)|F_n)$$
$$|n<t_1<\ldots<t_{k-j}\} \tag{2}$$

and the righthand side is given by

$$\sum_{i=1}^{j-1} Y_{n_i} r(\tau_{n_i}) + \text{ess sup}\{E(\sum_{i=1}^{k-j+1} Y_{t_i} r(t_i)|F_n)|n<t_1<\ldots<t_{k-j+1}\} \tag{3}$$

For $v \geq 0$ we set

$$\gamma(0,v):=0$$

$$\gamma(j,v):=\sup\{E(\sum_{i=1}^{j} Y_{t_i} r(\tau_{t_i}))|\ t_1<\ldots<t_j$$
$$\text{stopping times such that } \tau_{t_1} \geq v\}. \tag{4}$$

A little reflection shows that the ess sup in (2) is equal to $\gamma(k-j,\tau_n)$, and the ess sup in (3) equals $\gamma(k-j+1,\tau_n)$. Thus we define the above k-stopping rule more neatly by setting $s_o:=0$ and, recursively,

$$s_{k-j}:=\inf\{n>s_{k-j-1}|Y_n r(\tau_n)+\gamma(j,\tau_n)\geq\gamma(j+1,\tau_n)\},$$
$$j=k-1,\ k-2,\ldots,0. \tag{5}$$

$\gamma(j,v)$ is the maximal expected reward which can be realized by selling j commodities, if sale is allowed only after time v. We observe that $\gamma(j,v)<\infty$ for all j,v, since

$$E(\sup_{n_1<\ldots<n_k} X_{n_1,\ldots,n_k}) \leq E(\sum_{j=1}^{\infty} Y_j\, r(\tau_j)) = \mu\lambda \int_0^{\infty} r(u)\, du < \infty.$$

<u>Theorem 1</u>. $(s_1,\ldots,s_k)$ is an optimal k-stopping rule.

<u>Proof</u>. Theorem 2 in Stadje (1985) states that $(s_1,\ldots,s_k)$ is optimal, if all $s_i$ are a.s. finite and if

$$E(\sup_{m\geq 1} E(\sup_{n_1<\ldots<n_k} X_{n_1,\ldots,n_k} | F_m)) < \infty \tag{6}$$

holds. Thus we have to check these conditions. (6) follows from

$$\begin{aligned}
&E(\sup_{m\geq 1} E(\sup_{n_1<\ldots<n_k} X_{n_1,\ldots,n_k} | F_m)) \\
&\leq E(\sup_{m\geq 1} [\sum_{j=1}^{m} Y_j\, r(\tau_j) + \sup_{m<n_1<\ldots<n_k} X_{n_1,\ldots,n_k}] | F_m) \\
&\leq E(\sum_{j=1}^{\infty} Y_j\, r(\tau_j)) + E(\sup_{m\geq 1} E(\sup_{m<n_1<\ldots<n_k} X_{n_1,\ldots,n_k} | F_m)) \\
&\leq \lambda\mu \int_0^{\infty} r(u)du + E(\sup_{m\geq 1} \lambda\mu \int_{\tau_m}^{\infty} r(u)du) \\
&\leq 2\lambda\mu \int_0^{\infty} r(u)du < \infty.
\end{aligned}$$

To see that $s_1$ is a.s. finite, we note the obvious inequalities

$$\gamma(k,t) \leq \gamma(1,t) + \gamma(k-1,t)$$

$$\gamma(1,t) \leq r(t)\gamma(1,0)$$

(for the second inequality (1") is needed). Thus $s_1 \leq s_1' := \inf\{n\geq 1 | Y_n \geq \gamma(1,0)\}$. Let $\eta := \text{ess sup } Y_1$. Since $r(u) \to 0$, as $u \to \infty$, there is an $u_0 > 0$ such that $r(u) < 1/2$ for all $u \geq u_0$. Hence if $\eta < \infty$, we obtain for each stopping time t,

$$E(Y_t\ r(t)) \leq E(Y_t\ 1_{\{\tau_1 < u_o\}}) + \frac{1}{2} E(Y_t\ 1_{\{\tau_1 \geq u_o\}})$$

$$\leq \eta\ P(\tau_1 < u_o) + \frac{1}{2} P(\tau_1 \geq u_o)$$

$$= (1 - \frac{1}{2} e^{-\lambda u_o})\eta$$

so that $\gamma(1,0) < \eta$. This inequality is trivially also true if $\eta = \infty$. Thus $P(Y_1 > \gamma(1,0)) > 0$, and, consequently, $s_1' < \infty$ a.s. Thus $P(s_1 < \infty)=1$. The a.s. finiteness of $s_2,\ldots,s_k$ follows by induction.

Let $x(j,t) := \gamma(j,t) - \gamma(j-1,t)$. $x(j,t)$ is the additional expected reward, if j instead of j-1 commodities are for sale after time t. Clearly we have for $j=1,\ldots,k-1$ and $k \in \mathbb{N}$

$$\gamma(k,t) = \sum_{j=1}^{k} x(j,t) \tag{7}$$

$$s_{k-j} = \inf\{n > s_{k-j-1} \mid Y_n r(\tau_n) \geq x(j+1,\tau_n)\}. \tag{8}$$

Further let $h(u) := E((Y_1 - u)^+)$. The essential properties of the functions $x(j,t)$ are summarized in the following Theorem.

<u>Theorem 2</u>. (a) For each j, $x(j,t)$ is continuous, piecewise differentiable on the joint continuity intervals of r and F and there satisfies the differential equation

$$x'(1,t) := \frac{d}{dt} x(1,t) = -\lambda r(t) h(x(1,t)/r(t))$$

$$x'(j,t) := \frac{d}{dt} x(j,t) = -\lambda r(t)[h(x(j,t)/r(t)) - h(x(j-1,t)/r(t))],\ j \geq 2.$$

We set $0/0 := 0$.

(b) $\lim_{t\to\infty} x(j,t) = 0$ for $j=1,2,\ldots$

(c) $(x(j,t))_{j\geq 1}$ is uniquely determined by (a) and (b).

(d) $x(j,t)$ is monotone decreasing with respect to j and to t.

<u>Proof</u>. For $v \geq 0$ let $(s_1(v),\ldots,s_k(v))$ be a k-stopping rule for which the supremum in (2) with j=k is attained. Then it is clear that we can choose $s_1(v)$ as follows:

$$s_1(v) = \inf\{n \geq 1 \mid Y_n r(\tau_n) \geq x(k,\tau_n),\ \tau_n \geq v\}.$$

Let $\alpha(v) := \tau_{s_1(v)}$. Given that $\alpha(v) = u$, the maximal expected reward for the k-stopping problem with stopping allowed at the earliest at time v is equal to

$$r(u)E(Y_1 \mid \alpha(v) = u) + \gamma(k-1,u)$$
$$= r(u)E(Y_1 \mid Y_1 \geq x(k,u)/r(u)) + \gamma(k-1,u)$$

so that

$$\gamma(k,v) = \int_v^\infty [r(u)E(Y_1 \mid Y_1 \geq x(k,u)/r(u)) + \gamma(k-1),u)]P(\alpha(v) \in du). \tag{9}$$

Next we determine the distribution of $\alpha(v)$. Let M and N be the number of the first and the last offer in $[v,u]$, respectively. Given that $N-M+1 = n$, the time instants $\tau_M,\ldots,\tau_{M+n-1}$ form an ordered sample of size n from the uniform distribution on $[v,u]^n$. This is a standard property of the Poisson process (Karlin (1969), ch. 9). Given that $N-M+1 = n$ and $(\tau_M,\ldots,\tau_{M+n-1}) = (y_1,\ldots,y_n) \in [v,u]^n$ the event $\{\alpha(v) > u\}$ occurs if $Y_M < x(k,y_1)/r(y_1),\ldots,Y_{M+n-1} < x(k,y_n)/r(y_n)$. As $\tau_1,\tau_2,\ldots,M,N$ are independent of $Y_1,Y_2,\ldots,$ this latter event has probability

$$\prod_{i=1}^n F\left(\frac{x(k,y_i)}{r(y_i)} - 0\right). \tag{10}$$

Therefore conditioning yields

$$P(\alpha(v) > u) = \sum_{n=0}^\infty \frac{[\lambda(u-v)]^n}{n!} e^{-\lambda(u-v)}$$
$$\int\cdots\int_{v \leq y_1 < \ldots < y_n \leq u} n!(u-v)^{-n} P(\alpha(v) > u \mid N-M+1 = n, (\tau_M,\ldots,\tau_{M+n-1})$$
$$=(y_1,\ldots,y_n))\, dy_1\ldots dy_n$$
$$= e^{-\lambda(u-v)} \sum_{n=0}^\infty \frac{\lambda^n}{n!} \int_v^u\cdots\int_v^u \prod_{i=1}^n F\left(\frac{x(k,y_i)}{r(y_i)} - 0\right) dy_1\ldots dy_n$$
$$= \exp\left\{-\lambda \int_v^u \left[1-F\left(\frac{x(k,s)}{r(s)}\right)\right]ds\right\}. \tag{11}$$

Let $\tilde{F}(s) := \int_s^\infty y\,dF(y)$. From (9) and (11) it follows that

$$\gamma(k,v) = \lambda \int_v^\infty \left[r(u)\tilde{F}\left(\frac{x(k,u)}{r(u)}\right) + \left(1-F\left(\frac{x(k,u)}{r(u)}\right)\right) \gamma(k-1,u)\right] \times$$

$$\times \exp\{-\lambda \int_v^u [1-F(\frac{x(k,s)}{r(s)})]\, ds\}\, du. \tag{12}$$

Since F and r are piecewise continuous, we can conclude recursively from (12) that all $\gamma(j,.)$ and hence all $x(j,.)$ are continuous and piecewise differentiable (and differentiable from both sides at the breakpoints). (a) can now be derived from (12) by an easy induction argument. (a) is proved.

To show (d) we first prove $x'(j,t) \leq 0$ for all $j \in \mathbb{N}$ and $t \geq 0$ by induction. For j=1 we have $x'(1,t) = -\lambda r(t) h(x(1,t)/r(t)) \leq 0$ by (a). If the assertion holds for some j, (12) entails

$$\begin{aligned}\gamma(j+1,t) &= \int_t^\infty [r(u)\, E(Y_1 | Y_1 \geq \frac{x(j+1,u)}{r(u)}) + \gamma(j,u)]\, P(\alpha(t) \in du) \\ &= \int_t^\infty [r(u)\, E(Y_1 | Y_1 \geq \frac{x(j+1,u)}{r(u)}) + \gamma(j-1,u)]\, P(\alpha(t) \in du) \\ &\quad + \int_t^\infty x(j,u)\, P(\alpha(t) \in du). \end{aligned} \tag{13}$$

By the induction hypothesis, the last integral on the righthand side is at most equal to $x(j,t)$. The first integral on the righthand side can be interpreted as the expected reward, if j commodities are for sale after time t, the first stopping is at time $\alpha(t)$ and thereafter an optimal procedure is used. Thus this integral does not exceed $\gamma(j,t)$. We obtain

$$\gamma(j+1,t) \leq \gamma(j,t) + x(j,t) \tag{14}$$

which implies that

$$x(j+1,t) \leq x(j,t). \tag{15}$$

By (15) and the monotonicity of h,

$$x'(j+1,t) = -\lambda r(t)[h(x(j+1,t)/r(t)) - h(x(j,t)/r(t))] \leq 0.$$

(d) is proved. Further we have

$$x(j,t) \leq x(1,t) \leq \lambda\mu \int_t^\infty r(u)du \to 0, \text{ as } t \to \infty \tag{16}$$

so that (b) holds. To show (c), we note that h is convex, decreasing, and $h'(0+) = 1$. Thus the function $H(t,x) := -\lambda r(t) h(x/r(t))$ is easily seen to satisfy a Lipschitz condition with respect to x; in fact, $|H(t,x) - H(t,\bar{x})| \leq \lambda |x - \bar{x}|$. Therefore the equation $x'(1,t) = -\lambda r(t)\, h(x(1,t)/r(t))$ determines its solution on each joint continuity interval of $r(t)$ and $F(t)$ up to an additive constant. Because of the global continuity of each solution two solutions only differ for some fixed constant. Therefore the condition $x(1,t) \to 0$, as $t \to \infty$, singles out exactly one solution. That (a) and (b) uniquely determine $x(j,t)$ for any j now follows by an easy induction.
This completes the proof.

## III. EXAMPLES

We now consider the special case $r(t) = 1_{[o,T)}(t)$ for some fixed $T>0$: T is the last time point of profitable selling, and up to T there is no discounting. Finally the discount function $r(t) = e^{-\alpha t}$ will be treated.

Theorem 3. Let $F(x) = 1-e^{-x/\mu}$, $x \geq 0$, be the exponential d.f. with mean $\mu$. Then we have

$$x(j,t) = \mu[\log(\sum_{i=0}^{j} \frac{[\lambda(T-t)]^i}{i!}) - \log(\sum_{i=0}^{j-1} \frac{[\lambda(T-t)]^i}{i!})], \quad j=1,2,... \tag{17}$$

Proof. In this case a short calculation shows that $h(u) = \mu e^{-u/\mu}$, $u \geq 0$. (a) and (b) of Theorem 2 take the form

$$x'(j,t) = -\lambda\mu(\exp\{-x(j,t)/\mu\} - \exp\{-x(j-1,t)/\mu\}) \tag{18}$$

$$x(j,T) = 0.$$

The solution of these recursive differential equations is given by (17).

Next we consider the case when $Y_1$ only attains two values a,b where $0<a<b$. Let $P(Y_1 = a) = p$, $P(Y_1 = b) = q = 1-p$, $p \in (0,1)$. Then $\mu = E(Y_1) = pa + qb$. We define $t_1, t_2$ by

$$t_1 := T - \lambda^{-1}\log\frac{\mu}{\mu-a} \tag{20}$$

$$t_2 := (q\lambda)^{-1}\log u, \tag{21}$$

where $u>0$ is the (uniquely determined) solution of the equation

$$(1 + \log\frac{\mu}{\mu-a})(\frac{\mu-a}{\mu})^q e^{-\lambda T} u^{1/q} = (\frac{\mu-a}{\mu})^{q+1} - p^{-1}e^{-\lambda qT}u. \tag{22}$$

It will be seen that $t_2 \leq t_1$.

We shall only describe the optimal k-stopping rules for k=1 and k=2. Clearly an offer $Y_n = b$ is always accepted. If only one commodity is to be sold, after time $t_1$ also an offer $Y_n = a$ is accepted. If there are still two commodities for sale, one accepts an offer $Y_n = a$ from time $t_2$ on:

$$s_1 = \inf\{n \geq 1 | Y_n = b \text{ or } Y_n = a, \tau_n \geq t_1\}, \text{ if } k=1 \tag{23}$$

$$\left.\begin{aligned} s_1 &= \inf\{n \geq 1 \mid Y_n = b \text{ or } Y_n = a,\ \tau_n \geq t_2\} \\ s_2 &= \inf\{n > s_1 \mid Y_n = b \text{ or } Y_n = a,\ \tau_n \geq t_1\} \end{aligned}\right\} \text{ if } k=2. \tag{24}$$

These results are consequences of the proof of the following Theorem, where we shall show that $x(j,t) \leq a$ iff $t \geq t_j$, $j=1,2$.

Theorem 4. Let $0<a<b<\infty$, $0<p<1$, $q=1-p$, $P(Y_1=a)=p$, $P(Y_1=b)=q$. Then

$$x(1,t) = \begin{cases} \mu(1-e^{-\lambda(T-t)}), & \text{if } t \geq t_1 \\ b-(b-a)e^{\lambda q(t-t_1)}, & \text{if } t < t_1 \end{cases} \tag{25}$$

$$x(2,t) = \begin{cases} \mu(1-[\lambda(T-t)+1]\, e^{-\lambda(T-t)}), & \text{if } t \geq t_1 \\ \mu[1-p^{-1}e^{\lambda q(t-t_1)} - (1+\log\frac{\mu}{\mu-a})\, e^{-\lambda(T-t)}], & \text{if } t_2 \leq t < t_1 \\ b-(b-a)[e^{\lambda q(t-t_2)} - q(t-t_2)\, e^{\lambda q(t-t_1)}], & \text{if } t < t_2. \end{cases} \tag{26}$$

Here $t_1$ and $t_2$ are given by (20 and (21).

Proof. We have

$$h(u) = \begin{cases} \mu-u, & \text{if } u \leq a \\ q(b-u), & \text{if } a<u\leq b \\ 0, & \text{if } b<u. \end{cases} \tag{27}$$

Thus on the set where $x(1,t) < a$,

$$x'(1,t) = -\lambda h(x(1,t)) = \lambda x(1,t) - \lambda\mu. \tag{28}$$

From (28) and the boundary condition $x(1,T) = 0$ we obtain $x(1,t) = \mu(1-e^{-\lambda(T-t)})$. This solution satisfies $x(1,t) < a$ iff $t > t_1$, where $t_1$ is defined by (20). The relations

$$x'(1,t) = -\lambda q(b-x(1,t)), \quad \text{if } a<x(1,t)<b \tag{29}$$

$$x(1,t_1) = a \tag{30}$$

then yield the second part of (25) after a little computation.

Now suppose that $t>t_1$, so that $x(1,t)<a$ and, by Theorem 2(d), $x(2,t)<a$, too. In this case the differential equation for $x(2,t)$ takes the form

$$x'(2,t) = -\lambda[h(x(2,t)) - h(x(1,t))]$$

$$= \lambda x(2,t) - \lambda\mu + \lambda\mu\, e^{-\lambda(T-t)}, \quad t > t_1 \tag{31}$$

$$x(2,T) = 0, \tag{32}$$

from which the first line of (26) follows. Next let $t \le t_1$ and $x(2,t) < a$. In this case our equation is given by

$$x'(2,t) = \lambda x(2,t) - \lambda\mu + \lambda q(b-a)\, e^{\lambda q(t-t_1)} \tag{33}$$

$$x(2,t_1) = \mu(1-[\lambda(T-t_1)+1]\, e^{-\lambda(T-t_1)}) = a - (\mu-a)\log\frac{\mu}{\mu-a} \tag{34}$$

and has the solution

$$x(2,t) = \mu[1 - p^{-1} e^{\lambda q(t-t_1)} - (1+\log\frac{\mu}{\mu-a})\, e^{-\lambda(T-t)}]. \tag{35}$$

The condition $x(2,t) \le a$ holds iff $t \ge t_2$ with $t_2$ given by (21) and (22). Finally assume that $t < t_1$. Then

$$\begin{aligned} x'(2,t) &= -\lambda[q(b-x(2,t)) - q(b-x(1,t))] \\ &= \lambda q x(2,t) - \lambda q b + q(b-a)\, e^{\lambda q(t-t_1)} \end{aligned} \tag{36}$$

$$x(2,t_2) = a. \tag{37}$$

Solving (36) and (37) yields the last line of (26), as is easily checked.

Now let all $Y_j$ be uniformly distributed on (0,1). Then we have $h(u) = (1-u)^2/2$, $u \in [0,1]$. The computation is only carried out for k=1 and k=2, because for $k \ge 3$ the originitating Riccati differential equations seem to be intractable (in closed form). In what follows, it is convenient to consider $y_j(t) := x(j,T-t)$.

<u>Theorem 5</u>. Let $F(x) = x$, $0 \le x \le 1$. Then we have

$$y_1(t) = \lambda t/(\lambda t + 2) \tag{38}$$

$$y_2(t) = 1 - (\lambda t + 2)^{-1} \frac{(1+\sqrt{5})^2 (1+\frac{\lambda t}{2})^{\sqrt{5}} - (1-\sqrt{5})^2}{(1+\sqrt{5})(1+\frac{\lambda t}{2})^{\sqrt{5}} - 1+\sqrt{5}}. \tag{39}$$

<u>Proof</u>. The recursive equation for $y_1, y_2, \ldots$ is given by

$$y_j'(t) = \lambda[h(y_j(t)) - h(y_{j-1}(t))] \tag{40}$$

$$= \frac{\lambda}{2}[(1-y_j(t))^2 - (1-y_{j-1}(t))^2], \; j\geq 1$$

$$y_j(0) = 0, \; j\geq 1. \tag{41}$$

For j=1 it follows that $y_1'(t) = \frac{\lambda}{2}(1-y_1(t))^2$ and $y_1(0)=0$, so that $y_1(t) = \lambda t/(\lambda t+2)$. For j=2 we obtain the Riccati equation

$$y_2'(t) = \frac{\lambda}{2}[(1-y_2(t))^2 - (2/(\lambda t+2))^2] \tag{42}$$

$$y_2(0) = 0 \tag{43}$$

which happens to be solvable in closed form. The substitution $z(t) := (\frac{\lambda t}{2}+1)(1-y_2(t))$ yields

$$z'(t) = -\frac{\lambda}{2}(z(t)^2 - z(t)-1)/(\frac{\lambda t}{2}+1) \tag{44}$$

$$z(0) = 0. \tag{45}$$

Let $z_1 = \frac{1}{2}(1-\sqrt{5})$ and $z_2 = \frac{1}{2}(1+\sqrt{5})$. Then, by (44),

$$-\log(\frac{\lambda t}{2}+1) + C = \int \frac{z'(t)\,dt}{(z-z_1)(z-z_2)}$$

$$= 5^{-1/2} \log \left| \frac{z(t)-z_2}{z(t)-z_1} \right| . \tag{46}$$

We observe that $z(t)$ cannot attain the values $z_1, z_2$. Since $z(0)=1\in(z_1,z_2)$, we conclude that $z(t)\in(z_1,z_2)$ for all $t\geq 0$. Moreover setting t=0 gives

$$C = 5^{-1/2} \log \frac{-z_1}{z_2} . \tag{47}$$

Thus, by (46) and (47),

$$\frac{z_2-z(t)}{z(t)-z_1} = -\frac{z_1}{z_2} (\frac{\lambda t}{2}+1)^{-\sqrt{5}} \tag{48}$$

Solving for $z(t)$ now easily yields (39).

Let us finally look at the case $r(t) = e^{-\alpha t}$ for some $\alpha > 0$. By the lack of memory of the Poisson process and the independence of $(Y_n)_{n\geq 1}$ and $(\tau_n)_{n\geq 1}$,

$$\gamma(j,u) = \text{ess sup}\{E(\sum_{i=1}^{j} Y_{t_i} r(\tau_{t_i})) \mid t_1 < \ldots < t_j,\ \tau_{t_1} \geq u\}$$

$$= e^{-\alpha u} \text{ ess sup}\{E(\sum_{i=1}^{j} Y_{t_i} \exp(-\alpha(\tau_{t_i} - u)) \mid$$

$$t_1 < \ldots < t_j,\ \tau_{t_1} - v \geq 0\}$$

$$= e^{-\alpha u}\ \gamma(j,0).$$

Hence, setting $x_j := x(j,0)$ we obtain

$$x(j,t) = x_j\ e^{-\alpha t}, \qquad t \geq 0. \tag{49}$$

By Theorem 2, the constants $x_j$ satisfy $x_1 > x_2 > x_3 > \ldots > 0$ and

$$-\alpha x_j\ e^{-\alpha t} = x'(j,t) = -\lambda e^{-\alpha t}[h(x_j) - h(x_{j-1})],\ j \geq 2$$

$$-\alpha x_1\ e^{-\alpha t} = x'(1,t) = -\lambda e^{-\alpha t}\ h(x_1).$$

Thus $x_1$ is a positive solution of

$$h(x) = (\alpha/\lambda)x \tag{50}$$

and, for $j \geq 2$, $x_j$ is given recursively as a positive solution of

$$h(x) = (\alpha/\lambda)x + h(x_{j-1}). \tag{51}$$

The positive solution of (50) or (51) is uniquely determined, because h is convex, $h(0) = \mu > 0$, and h is strictly decreasing on the interval $\{x \geq 0 \mid h(x) > 0\}$.

Special cases. (a) For the exponential distribution ($F(x) = 1-e^{-x/\mu}$) $x_1$ is given by

$$e^{-x_1/\mu} = \alpha x_1/\lambda\mu$$

and $x_j$ satisfies

$$e^{-x_j/\mu} = (\alpha x_j/\lambda\mu) + e^{-x_{j-1}/\mu}$$

(b) For the two point distribution of Theorem 4 h is given by (27), and the figure below shows the construction of the $x_j$.

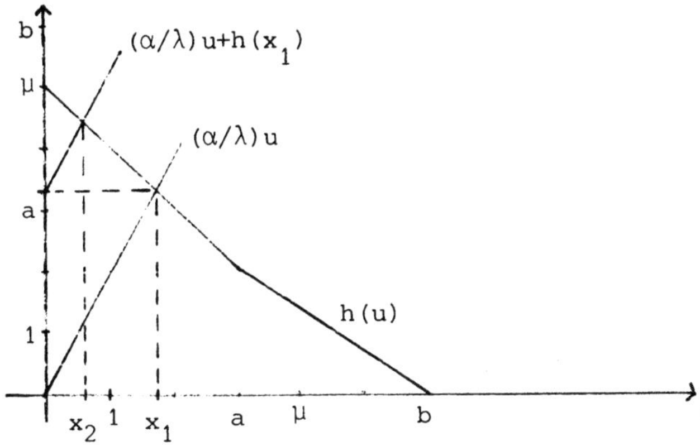


Construction of $x_1, x_2$ for a=3, b=6, p=1/3, λ=1, α=2.

(c) For the uniform distribution on (0,1) we have

$$x_1 = \frac{\lambda}{\alpha}(1-x_1)^2,$$
$$x_j = \frac{\lambda}{\alpha}[(1-x_j)^2 - (1-x_{j-1})^2], \quad j \geq 2$$

so that

$$x_1 = 1 - (2\lambda)^{-1}\alpha(\sqrt{1+2\alpha^{-1}\lambda} - 1)$$
$$x_j = 1 - (2\lambda)^{-1}\alpha(\sqrt{1+2\alpha^{-1}\lambda + 4\alpha^{-1}\lambda^2(1-x_{j-1})^2} - 1), \quad j \geq 2.$$

## REFERENCES

1. F.T. Bruss (1986). On an optimal selection problem of Cowan and Zabczyk. Preprint.

2. Y.S. Chow, H. Robbins and D. Siegmund (1971). Great Expectations: The Theory of Optimal Stopping. Boston, Houghton, Mifflin Co.

3. Z. Cieselski and J. Zabczyk (1979). Note on a selection problem. In: Probab. Theory, Banach Center Publ. 5, PWN, Warsaw.

4. R. Cowan and J. Zabczyk (1979). An optimal selection problem associated with the Poisson process. Theory Prob. Appl. 23, 584-592.

5. G. Elfving (1967). A persistancy problem connected with a point process. J. Appl. Prob. 4, 77-89.

6. P.R. Freeman (1983). The secretary problem and its extensions. Internat. Statist. Rev. 51, 189-206.

7. D.P. Gaver (1976). Random record models. J. Appl. Prob. 13, 538-547.

8. M. Henke (1970a). Optimale Stopp- und Auswahlregeln für eine Klasse stochastischer Entscheidungsprozesse. Op. Res. Verfahren VII, 83-121.

9. M. Henke (1970b). Sequentielle Auswahlprobleme bei Unsicherheit. Meisenheim, Anton Hain Verlag.

10. S. Karlin (1969). A First Course in Stochastic Processes. New York, Acad. Press.

11. D.P. Kennedy (1986). Prophet-type inequalities for multichoice optimal stopping. Preprint.

12. T.F. Móri (1984). The random secretary problem with multiple choice. Ann. Univ. Sci. Bud. R. Eötvös Nom., Sect. Comput. 5, 91-102.

13. J.D. Petrucelli (1982). Full information best-choice problems with recall of observations and uncertainty of selection depending on the observation. Adv. Appl. Prob. 14, 340-358.

14. M. Sakaguchi (1976). Optimal stopping problems for randomly arriving offers. Math. Japon. 21, 201-217.

15. D. Siegmund (1967). Some problems in the theory of optimal stopping. Ann. Math. Statist. 38, 1627-1640.

16. W. Stadje (1985). On multiple stopping rules. Optimization 16, 401-418.

17. M. Tamaki (1979). OLA policy and the best choice problem with random number of objects. Math. Japon. 24, 451-457.

# A STATISTICAL APPROACH TO RELAXATION IN GLASSY MATERIALS

Karina Weron
Institute of Physics
Technical University of Wroclaw
50-370 Wroclaw, Poland

Aleksander Weron
Institute of Mathematics
Technical University of Wroclaw
50-370 Wroclaw, Poland

ABSTRACT. Statistical aspects of the relaxation model in amorphous materials are studied. After a brief overview of the physical literature on the subject, it is proved that the relaxation rate distribution is described by completely asymmetric p-stable distributions with the index of stability $0 < p < 1$ and the skewness parameter $\beta = 1$. This gives a rigorous approach to the stretched exponential form of the relaxation function. It is also shown how adaptive estimation of the parameters of stable distribution can be used in the study of relaxation phenomena.

## 1. INTRODUCTION

Relaxation in amorphous materials, like glassy materials or viscous liquids, is the time-dependent change in any macroscopic material property (density, enthalpy, optical properties or structure factor) following a perturbation (change in temperature, stress, electric or magnetic field). In amorphous polymers this process is known as "aging". Relaxation phenomena are of great technological importance. For instance, it may be important for optical fibers, especially if they are fabricated from halide glasses, which tend to have low glass transition temperatures. Relaxation of the fiber could change its shape (at least its cross section) and thus alter the propagation characteristic in material [1]. An interesting technological consequence of relaxation is the change of calibration of mercury thermometers with time, due to a change in the shape of the thermometer bulb [2].

It has been observed a considerable progress in understanding the relaxation phenomena in the past 20 years, however, theoretical advances are still not satisfactory. The purpose of this paper is to present a statistical approach to the relaxation in amorphous materials which is based on a microscopic model of the material. The basic physical



*P. Bauer et al. (eds.), Mathematical Statistics and Probability Theory, Vol. B, 245–254.*

idea is that the main features of such materials are due to heterogeneous nature of the microscopic structure. By the heterogeneity is meant the fact that structure fluctuates from point to point, which leads to a distribution of relaxation time.

The experimental features are observed as a broad distribution of relaxation times and the result of a typical experiment is commonly presented in the form of a relaxation function $R(t)$, as a function of time t. It is usually found [3-6] that

$$R(t) = \exp\{-(t/\tau_e)^p\}, \qquad 0 < p < 1 \tag{1.1}$$

with p and $\tau_e$ constants for a given material, where $\tau_e$ is an effective relaxation time. This form of relaxation function is termed the Williams-Watts function. Equation (1.1) has been shown to be good description of mechanical and dielectric relaxation [7-10], and of quasielastic light-scattering experiments [11]. This is in contrast to the conventional Debye exponential form

$$R(t) = \exp(-t/\tau_o), \tag{1.2}$$

observed especially for materials composed of simple molecules, where $\tau_o$ is termed the primitive relaxation time.

It is a striking fact that, despite of the variety of materials used and of the experimental techniques employed, the relaxation function is universal. The recent interest in relaxation problem provides a number of models [6,12-20] explaining the universality of formula (1.1). A connection of the function $R(t)$ with the well known in probability theory class of p-stable distributions [21-25] has been investigated in [16-20].

By analogy with the Sherr-Montroll theory [26] of charge transport in complex amorphous materials, the dielectric relaxation described by the Williams-Watts function (1.1) is interpreted by Montroll and Bendler [17] as the survival probability of a frozen segment in swarm of hopping defects with a stable waiting-time distribution $At^{-p}$ for defect motion. The exponent p is the fractal dimension of a hierarchical scaling set of defect hopping times. Shlesinger and Montroll [16] consider the flux of many defects,and a survival probability of the form above results at long times in three dimensions for a swarm of nonbiased defects hopping to a single dipole. Shlesinger [19] introduced the fractal time stochastic process and derived for the fractal time defect-diffusion model the Williams-Watts relaxation law. The argument of the exponent is related to the number of distinct sites visited after a time t by a random walker.

All these results are based on the continuous time random walk approach and the behaviour of the waiting-time distribution at long times. A different approach has been proposed in [20], constituting the basis for the subsequent sections of this paper. Many other scaling relations for complex systems in physical sciences involve non-integer exponents, similarly to the Fourier transforms of p-stable distributions. Consequently, there are many results (e.g., in statistical physics, electronics, quantum mechanics) where p-stable distributions are invoked [23,25].

A connection between the relaxation function and the class of p-stable distributions is not only a coincident analogy but seems to capture the essential statistical nature of relaxation in complex materials. From this point of view p-stable distributions form a class of universal laws in the sense that they describe completely the collective, i.e., macroscopic behaviour of a complex system expressed by any normalized sum of independent identically distributed (i.i.d.) quantities. Consequently, as an answer to the question: Why does equation (1.1) hold for so many materials? - we shall deduce it from the general limit theorem.

## 2. THE RELAXATION FUNCTION

The stable distributions are the only limiting distributions of normalized sums of i.i.d. random variables. P. Lévy [21] discovered these distributions and also computed their characteristic functions. There exist four real constants $p, \beta, \sigma$ and $m$ with $0 < p \leqq 2$, $-1 \leqq \beta \leqq 1$ and $\sigma \geqq 0$ so that the characteristic function $C(t)$ of a stable distribution has the form:

$$C(t) = C(t;p,\beta,\sigma,m) = \exp(imt - \sigma|t|^p[1-i\beta\,\mathrm{sgn}(t)w(t;p)]), \tag{2.0}$$

where

$$w(t;p) = \begin{cases} \tan(\pi p/2) & p \neq 1 \\ -\frac{2}{\pi}\log|t| & p = 1 \end{cases}$$

and $\mathrm{sgn}(t) = 1$, $0$ or $-1$ according to $t > 0$, $t = 0$ or $t < 0$. Here $p$ is called the index of stability (in which case we call the distribution p-stable), $\beta$ is a skewness parameter, $m$ is a location and $\sigma$ is a scale parameter. The skewness parameter $\beta$ gives a measure of how much of the Lévy-Khintchine jump measure is placed on the positive and negative half lines. For example, $\beta = 1$, $-1$ or $0$ according as the measure is concentrated on the right half-line, the left half-line, or is symmetric. For the relevant references see [23,24,25].

Proposition. In a complex amorphous material a general relaxation function defined in terms of exponential relaxations, i.e.,

$$R(t) = \int_0^\infty \exp(-t/\tau)\rho(\tau)d\tau,$$

where $\rho(\tau)$ is the density of the distribution of relaxation times, has the following form

$$R(t) = \exp(-mt/\tau_o - [\sigma/\cos(\pi p/2)](t/\tau_o)^p),$$

where $0 < p < 1$, $\sigma \geqq 0$ and $m \geqq 0$.

Proof. In the literature one finds several definitions of the relaxation time distribution. It is common [17,27,28] to attempt to interpret general relaxation in terms of a superposition of exponential relaxations, writing

$$R(t) = \int_0^\infty \exp(-t/\tau)\rho(\tau)d\tau, \tag{2.1}$$

where $\rho(\tau)$ is the density of the relaxation time distribution or the density of a random variable T - the relaxation time, i.e., $\rho(\tau)d\tau$ is a measure of the contribution to the process of modes with relaxation times between $\tau$, $\tau + d\tau$. If $U = \tau_o/T$, where $\tau_o$ is a single relevant relaxation time associated with the Debye relaxation, then U can be considered as a random variable describing the relaxation rate and can be interpretable as dimensionless time. If one put $s = t/\tau_o$ in formula (2.1) then

$$R(\tau_o s) = \int_0^\infty (\tau_o u^{-2})\exp(-su)\rho(\tau_o/u)du = \int_0^\infty \exp(-su)r(u)du,$$

where

$$r(u) = \tau_o u^{-2}\rho(\tau_o/u)$$

is the density of the relaxation rate distribution.

Since our approach is microscopically arbitrary according to the heterogeneous nature of the microscopic structure, one may consider the random quantities $U_k = \tau_o/T_k$ as the possible relaxation rates of k-th element in a given complex amorphous material. The index k indicates the number of an element in a large system describing the given complex amorphous material. Without loss of generality one can assume that the random variables $U_k$ are mutually independent and identically distributed. The collective, i.e., macroscopic behaviour of the system is then completely described by the sums $\sum_{k=1}^{n} (U_k/a_n - b_n)$, where $a_n > 0$ and $b_n$ are normalization constants.

By the general limit theorem [22-25] the limit distribution $r(u)du$ of the macroscopic relaxation rate belongs to the class of p-stable distributions, $0 < p \leqq 2$. Since from the definition the relaxation rates are non-negative, $r(u)$ has to have non-negative support. It is well known [22] that this can happen only for a completely asymmetric p-stable distribution with $0 < p < 1$ and the skewness parameter $\beta = 1$. Let us denote its density by $f(u;p,1,\sigma,m)$. Hence

$$R(\tau_o s) = \int_0^\infty \exp(-su)\ f(u;p,1,\sigma,m)du. \tag{2.2}$$

Consequently, $R(\tau_o s)$ is the Laplace transform of a p-stable density $f(u;p,1,\sigma,m)$. Let us recall that the Fourier transform of $f(u;p,1,\sigma,m)$ can be written as

$$\begin{aligned} C(s) &= \exp\{ims - \sigma s^p[1 - i\tan(\pi p/2)]\} = \\ &= \exp\{ims - [\sigma/\cos(\pi p/2)](-is)^p\}. \end{aligned} \tag{2.3}$$

Using the relation between Laplace and Fourier transforms for this same non-negative function $R(s) = C(is)$ we get

$$R(\tau_o s) = \exp\{-ms - [\sigma/\cos(\pi p/2)]s^p\}, \tag{2.4}$$

and finally

$$R(t) = \exp\{-mt/\tau_o - [\sigma/\cos(\pi p/2)](t/\tau_o)^p\}. \tag{2.5}$$

□

<u>Corollary</u>. As special cases we can deduce from (2.5) the Williams-Watts form (1.1) and the Debye form (1.2) of the relaxation function.

<u>Proof</u>. One can choose the location parameter $m = 0$ and $\sigma > 0$, then

$$R(t) = \exp(-(t/\tau_e)^p) , \qquad 0 < p < 1$$

with $\quad \tau_e = \tau_o^p/\sigma\cdot\cos(\pi p/2)$,

has the Williams-Watts form.

The second possibility is to choose $m > 0$ and $\sigma = 0$, then

$$R(t) = \exp(mt/\tau_o) = \exp(-t/\tilde{\tau}_o),$$

has the conventional Debye form with $\tilde{\tau}_o = \tau_o/m$.

□

Remark. Let us observe that the weight function $\rho(\tau)$ in (2.1) is expressible in terms of the stable density, namely

$$\rho(\tau) = \tau_o^{-1} u^2 \; f(u;p,1,\sigma,m).$$

Hence, one can employ the asymptotic formulas for stable densities to derive the asymptotic behaviour of the density of the relaxation time distribution $\rho(\tau)$ for $\tau \to 0$ and for $\tau \to \infty$.

## 3. TESTING FOR THE SHAPE OF THE RELAXATION FUNCTION

As we know from the previous section, the relaxation rates are described by a completely asymmetric p-stable distribution with the density $f(u;p,1,\sigma,m)$. In oder to get full information on relaxation phenomena it is desirable to know how to estimate the parameters: $0 < p < 1, \sigma \geqq 0$ and $m \geqq 0$. Many authors tried to fit the Williams-Watts function to experimental data, see [3-13] and reference therein. It turns out that the parameter p values generally range from 0.3 to 0.8, but there are also exceptions with smaller p values of the order of 0.2. For example, the data of Ishida and Yamafugi, cf. [17], on polyvinylacetate at 62.5°C were identified with $p = 0.56$.

Using the proposition proved in Section 2 we would like to use the adaptive estimation of the parameters of stable densities [29,30] in oder to estimate stretched exponential form of the relaxation function (1.1). The assumption of stability is usually motivated by the hypothesis that the random quantity in question is the sum of a large number of i.i.d. random variables and the observation that this sum is often dominated by one of the summands, a property of distributions lacking variance.

Let $U_1,\ldots,U_n$ be i.i.d. random variables describing possible relaxation rates of a given complex amorphous material with common characteristic function $C(t)$ of the form (2.0). Let

$$C_n(t) = 1/n \sum_{j=1}^{n} e^{itU_j}, \qquad (3.1)$$

be the empirical characteristic function of this sample. The idea is to testing the hypothesis

$$H : C(t) = \exp(imt - \sigma|t|^p[1 - i\,\mathrm{sgn}(t)w(t;p)]) \qquad (3.2)$$

for three parameters: $0 < p < 1$, $\sigma \geqq 0$ and $m \geqq 0$.

Following S. Csörgö [29] let two values of t, $t_1$ and $t_2$ be given such that $t_1 \neq 0$, $t_2 \neq 0$ and $|t_1| \neq |t_2|$. Then from the equation $C_o(t_k) = \exp(-\sigma|t_k|^p)$ (i.e., assuming that $m = 0$ and $\beta = 0$) we have

$$|t_k|^p = |\log|C_o(t_k)||, \qquad \text{for } k = 1 \text{ and } 2.$$

Solving these two equations simultaneously one obtains

$$p(t_1,t_2) = \frac{\log|\log|C_o(t_1)|| - \log|\log|C_o(t_2)||}{\log|t_1/t_2|}. \qquad (3.3)$$

It is known,[30] that for a characteristic function $C_o(t)$ the hypothesis $H_o : C_o(t) = \exp(-\sigma|t|^p)$ holds if and only if $C_o(t)$ belongs to a non-lattice distribution and the function $p(t_1,t_2)$ is constant. We may ignore the scale parameter $\sigma$ in the length of $C_o$ and use simplest ad hoc estimator

$$\hat{\sigma}_n = |\log|C_n(1)||. \qquad (3.4)$$

Consequently, the test will be based on the functional estimator of the parameter p

$$\hat{p}_n(t) = \hat{p}_n(t,t_2) = \frac{\log|\log|C_n(t)|| - \log|\log|C_n(t_2)||}{\log|t/t_2|}, \qquad (3.5)$$

where $C_o(\cdot)$ is replaced by $C_n(\cdot)$ and $t_1$ by t. Here $t_2$ is an arbitrarily fixed positive number.

Choosing $0 < t < t_2 < t^*$, where $t^*$ is the smallest positive root of the real part of $C(t)$ and $t_2$ is the same fixed number as above, we may express the estimator for m in the following form

$$\hat{m}_n(t) = \frac{m_n(t_2)t^{p-1} - m_n(t)t_2^{p-1}}{t^{p-1} - t_2^{p-1}}, \tag{3.6}$$

where $m_n(t) = t^{-1}\arg C_n(t)$.

According to the consideration in the previous section, our task in testing the shape of the relaxation function $R(t)$ is testing for the asymptotic constancy of the random continuous sheet $\hat{p}_n(t) + \hat{m}_n(s)$. This would lead to two-variate maximization and minimization problems. However, S. Csörgö [30] showed that one can consider the asymptotic constancy of the random continuous univariate function $\hat{p}_n(t) + \hat{m}_n(t)$ only. Then theorem 2.4.2 in [29] gives the necessary weak convergence result for

$$n^{1/2}(\hat{p}_n(t) + \hat{m}_n(t) - p - m),$$

and usual machinery leading to a test statistic works well, [29,30].

## 4. CONCLUDING REMARKS

There are several approaches to explaining non-Debye relaxation behaviour in amorphous materials [3-20]. The statistical approach interprets the stretched exponential form of relaxation function in these materials in terms of a superposition of exponentially relaxing processes which then leads to a distribution of relaxation times. In section 2 we derived the Williams-Watts formula directly from the general limit theorem. The application of this generalized central limit theorem gives that the distribution of the properly centralized and normalized sum of possible relaxation rates is approximately p-stable provided, of course, that the underlying distribution of the summands belongs to the domain of attraction of a stable law. This is similar to the Mandelbrot hypothesis [31] which gained great popularity as a model of various economic distributions, such as income and stock price. The most important consequence is that the theory of stable distributions can be directly employed to the study of relaxation behaviour in glassy materials.

While normality has enormus literature, publications on testing for stability are very scarce. Beginning with the pioneering works of Mandelbrot two decades ago the recent overview of Csörgö [29] contains more than sixty references on testing for stability. In section 3 we show how adaptive estimation of the four parameters of stable distributions can be used in testing for the shape of the relaxation function. Although more remains to be done, we feel that it has been demonstrated that the statistical approach to relaxation phenomena is worthy of further investigation.

We do not propose that the statistical approach is the physically more correct one, instead, we discuss the mathematical foundation and consequences of this approach which must be understood when physical interpretation is attached to the distribution functions underlying the relaxation. We would like to conclude the paper by quotation from E. P. Wigner [32]: "The first point is that mathematical concepts turn up in entirely unexpected connections. Moreover, they often permit an unexpectedly close and accurate description of the phenomena in these connections. Secondly, just because of this circumstance, and because we do not understand the reasons of their usefulness, we cannot know whether a theory formulated in terms of mathematical concepts is uniquely appropriate". Let it be an excuse for the authors, if they were not sufficiently modest in some formulation of this paper.

## ACKNOWLEDGMENTS

The support of the Grant CPBP 01.02 is gratefully acknowledged.
A first version of this paper was written during the second author's visit at the Center for Stochastic Processes, Department of Statistics, University of North Carolina at Chapel Hill and was supported in part by the AFOSR Grant No.F 49620 82 C 0009.

## REFERENCES

[1] S.E.Miller, A.G.Cynoweth eds., *Optical Fiber Telecommunications*, Academic Press, New York 1979.

[2] R.J.Charles, *Glass Tech.* **12**,24 (1971).

[3] G.Williams, D.C.Watts, *Trans.Faraday Soc.* **66**,80 (1970).

[4] G.Williams, D.C.Watts, S.B.Dev, A.M.North, *Trans.Faraday Soc.* **67**, 1323 (1977).

[5] A.K.Jonsher, *Nature* **267**,673 (1977).

[6] K.L.Ngai, *Comments Solid State Phys.* **9**,127 (1979); **9**,141 (1980).

[7] N.G.McCrum, B.E.Read, G.Williams, *Anelastic and Dielectric Effects in Polymeric Solids*, Wiley,London 1967.

[8] L.C.E.Struik, *Physical Aging in Amorphous Polymers and Other Materials*, Elsevier, Amsterdam 1978.

[9] A.V.Lesikar, C.T.Moynihan, *J.Chem.Phys.* **73**,1932 (1980).

[10] S.M.Rekhson, O.U.Mazurin, *J.Am.Cer.Soc.* **57**,327 (1974).

[11] G.D.Paterson, *Adv.Polym.Sci.* **48**,125 (1983).

[12] K.L.Ngai, A.K.Rajagopal, R.W.Rendel, S.Teitler, *Phys.Rev.* B **28**,6073 (1983).

[13] T.V.Ramakrishnan, ed., *Non-Debye Relaxation in Condensed Matter*, *World Scientific*, Singapore 1984.

[14] S.A.Brawer, *J.Chem.Phys.* **81**,2 (1984).

[15] R.G.Palmer, D.L.Stein, E.Abrahams, P.W.Anderson, *Phys.Rev.Lett.* **54**,958 (1984).

[16] M.F.Shlesinger, E.W.Montroll, *Proc.Natl.Acad.Sci.USA* **81**,1280 (1984).

[17] E.W.Montroll, J.T.Bendler, *J.Stat.Phys.* **34**,129 (1984).

[18] J.T.Bendler, *J.Stat.Phys.* **36**,625 (1984).

[19] M.F.Shlesinger, *J.Stat.Phys.* **36**,639 (1984).

[20] K.Weron, *Acta Phys.Pol.* **A70**,529 (1986).

[21] P.Lêvy, *Bull.Soc.Math.France.* **52**,49 (1924).

[22] W.Feller, *An Introduction to Probability and Its Applications*, vol.2, Wiley, New York 1966.

[23] V.M.Zolotarev, *One-dimensional Stable Distributions*, (in Russian), Nauka, Moscow 1983.

[24] P.Hall, *Bull.London Math.Soc.* **13**,23 (1981).

[25] A.Weron, *Lecture Notes in Math.* **1080**,306,Springer-Verlag, Berlin 1984.

[26] H.Scher, E.W.Montroll, *Phys.Rev.* B**12**,2455 (1975).

[27] C.P.Lindsey, G.D.Paterson, *J.Chem.Phys.* **73**,3348 (1980).

[28] E.Helfand, *J.Chem.Phys.* **78**,1931 (1983).

[29] S.Csörgö, *Colloquia Math.Soc.J.Bolyai* **36**, P.Rêvêsz ed., 305, North-Holland,Amsterdam 1984.

[30] S.Csörgö, *Colloquia Math.Soc.J.Bolyai* **45**, K.Sarkadi ed., 31, North-Holland, Amsterdam 1986.

[31] B.Mandelbrot, *J.Business* **40**,393 (1967).

[32] E.P.Wigner, *Comm.Pure Appl.Math.* **13**,1 (1960).

# SUBJECT INDEX

This index contains key-words of the individual papers collected in this volume; key-words are quoted only once per article.

# Mathematical Statistics and Probability Theory

## Volume A: Theoretical Aspects

*Proceedings of the 6th Pannonian Symposium on Mathematical Statistics, Bad Tatzmannsdorf, Austria, September 14–20, 1986*

edited by


M. L. PURI
*Indiana University, Bloomington, U.S.A.*

P. RÉVÉSZ
*Technical University, Vienna, Austria*

and

W. WERTZ
*Technical University, Vienna, Austria*


The Sixth Pannonian Symposium on Mathematical Statistics was held at Bad Tatzmannsdorf, Austria, September 14–20, 1986. More than 100 contributions were presented embracing a wide range of topics including probability theory, stochastic processes, foundations of statistics, decision theory and statistical methods.
A refereed selection of these contributions emphasizing the development of statistical and probabilistic methods appears in this volume. These cover four major topics: probability and stochastic processes, testing hypotheses, estimation, and applications.
A companion volume includes primarily papers on the foundations of statistics and probability theory.


ISBN 90–277–2580–2

CONTENTS OF VOLUME A

MIX
Papier aus verantwortungsvollen Quellen
Paper from responsible sources
FSC® C105338

If you have any concerns about our products,
you can contact us on
**ProductSafety@springernature.com**

In case Publisher is established outside the EU,
the EU authorized representative is:
**Springer Nature Customer Service Center GmbH**
**Europaplatz 3, 69115 Heidelberg, Germany**

Printed by Libri Plureos GmbH
in Hamburg, Germany